BOSTON STUDIES IN THE PHILOSOPHY OF SCIENCE

VOLUME L

PARANOIA: A STUDY IN DIAGNOSIS

VOLUME 102

BOSTON STUDIES IN THE PHILOSOPHY OF SCIENCE

EDITED BY ROBERT S. COHEN AND MARX W. WARTOFSKY

VOLUME L

YEHUDA FRIED AND JOSEPH AGASSI

PARANOIA: A STUDY IN DIAGNOSIS

D. REIDEL PUBLISHING COMPANY

DORDRECHT-HOLLAND / BOSTON-U.S.A.

Library of Congress Cataloging in Publication Data

Fried, Yehuda, 1929 –
Paranoia: a study in diagnosis.

(Boston studies in the philosophy of science; v. 50)
(Synthese library; v. 102)
Bibliography: p.
Includes indexes.
1. Paranoia. 2. Mental illness – Diagnosis.
3. Psychology, Pathological – Classification. I.
Agassi, Joseph, joint author. II Title. III. Series.
Q174.B67 vol. 50 [RC520] 50ls [616.8′97′075] 76 – 21816
ISBN 90 – 277 – 0704 – 9
ISBN 90 – 277 – 0705 – 7 pbk.

Published by D. Reidel Publishing Company,
P.O. Box 17, Dordrecht, Holland

Sold and distributed in the U.S.A., Canada and Mexico
by D. Reidel Publishing Company, Inc.
Lincoln Building, 160 Old Derby Street, Hingham,
Mass. 02043, U.S.A.

Printed in The Netherlands

"He who sees a madman without being moved by his state,
or who looks at him only for amusement,
is a moral monster."

J. DAQUIN,
Philosophie de la folie
Chambéry, 1791

EDITORIAL PREFACE

There is a curious parallel between the philosophy of science and psychiatric theory. The so-called demarcation question, which has exercised philosophers of science over the last decades, posed the problem of distinguishing science proper from non-science – in particular, from metaphysics, from pseudo-science, from the non-rational or irrational, or from the untestable or the empirically meaningless. In psychiatric theory, the demarcation question appears as a problem of distinguishing the sane from the insane, the well from the mentally ill. The parallelism is interesting when the criteria for what *fails* to be scientific are seen to be congruent with the criteria which define those psychoses which are marked by cognitive failure.

In this book Dr Yehuda Fried and Professor Joseph Agassi – a practicing psychiatrist and a philosopher of science, respectively – focus on an extreme case of psychosis – paranoia – as an essentially intellectual disorder: that is, as one in which there is a systematic and chronic delusion which is sustained by logical means. They write: "Paranoia is an extreme case by the very fact that paranoia is by definition a quirk of the intellectual apparatus, a logical delusion." (p. 2.)

Their attempt, therefore, is to use the analysis of the extreme case as an instrument to probe the whole classificatory scheme of psychiatric disorders, to test the specific theories of demarcation between mental illness and mental health of such major theorists as Jackson, Freud, Bleuler, Ey, Piaget, Foucault, R. D. Laing and Szasz, and the theories of the nature of scientific rationality of such thinkers as Bacon, Meyerson, Popper and Lakatos. Thus, Fried and Agassi are concerned with such questions as: When is a religious fanatic a mere straying soul, and when is he a psychotic seeking refuge in a crazy notion? Why are so many madmen so very intelligent, perceptive, imaginative, persistent, engaging? Why do we sometimes take as signs of mental illness the very same characteristics which at other times we so appreciate in young students and investigators? What distinguishes the crank from the nut, and both from the original thinker?

The context for the problem of classification is not simply a scientific or a clinical one, in psychiatry. It is a profoundly humanistic context, which concerns the way in which mental patients are to be characterized. Treating the mental patient as a human being requires an effort – one which began only in the eighteenth century, and is far from conclusion. Humanizing mental illness began with Freud – whose theory put all human beings on a continuum between the utterly sane and the utterly insane. This humanization has had some strange results. From the view that we are all more or less crazy, we have arrived at the view, expressed by some psychiatrists (among them, anti-psychiatrists) who deny the very existence of mental illness, and who eschew the distinction between normal and abnormal.

Fried and Agassi propose that there are some serious philosophical as well as clinical-diagnostic problems here. If we start from Freud's continuum – moving from mild through severe neurosis (as mild or severe distortion of reality), to psychosis (which distorts the patient's view beyond recognition) – then it would seem to follow that those who are incarcerated or hospitalized because they cannot cope are psychotic; but in fact, many neurotics are hospitalized, and some psychotics are not. The difficulty arises, on Fried and Agassi's view, because Freud's theory admits no clear demarcation between those ideas which are just plain crazy and those which are simply far out. And indeed, on their analysis, Freud's view can hardly allow for such a demarcation, since it pictures neurosis as mild distortion and psychosis as serious distortion of the truth. It therefore seems to suggest that all madness is error. But then, the serious problem arises, that any clinical determination of mental illness would seem to depend *either* on settling the hairy philosophical problem of the criterion of truth, *or* failing that, simply accepting common and received opinion as true without question.

Fried and Agassi set the classification problem into a variety of contexts: psychological of course; but also social, methodological and metaphysical; and they explore the bearing of all these contexts on the problems of diagnostic classification. They do not claim to go further – neither to questions of cause or etiology, on the one hand, nor to questions of therapy, on the other. Fried and Agassi try to look at mental illness diagnostically, without searching for its cause. They see hysteria as emotional fixation. Paranoia by contrast is seen as an intellectual fixation. They suggest further that hysteria is the paradigm for all neurosis, and paranoia the paradigm for all psychosis.

We cannot foretell how the psychiatric profession will respond to all this, but we see here in their theory of the madman as the arrested thinker the application of their views of the thinker: thinkers build tentative intellectual frameworks within which they develop problems and solutions. Arrested thinkers, they say, may be arrested by anxiety but nevertheless, can be understood in terms of a theory of cognitive function, which includes cognitive pathologies. This is why their work is on the interface of philosophy and psychiatry.

Center for Philosophy and History of Science ROBERT S. COHEN
Boston University MARX W. WARTOFSKY
November 1976

TABLE OF CONTENTS

FOREWORD

The writers of the present study come from different backgrounds but share interests and appraisals. We are both, in particular, unhappy about the present state of the art and take it to be rooted in a theoretical defect. We found ourselves peculiar in that we were particularly unhappy about two specific points. First, we took the paradoxes of paranoia seriously, and considered the fact that they were still unresolved as the symptom for that defect in contemporary theory; we assumed that the paradoxes of paranoia were not pursued by other writers precisely because they soon lead their students out of the contemporary framework. Second, we share deep concern over the fact that the distinction between neurosis and psychosis is so important, so natural, yet so poorly articulated and therefore so easy to get confused about. Finally, connecting these two points, and deeply appreciating Freud's view of psychopathology in everyday life, we felt that a proper demarcation between neurotic and psychotic should enable us to show instances, not only of (mild) neurosis, but also of (mild) psychosis in everyday life. The very existence of ambulatory paranoia and of similarities between paranoics and geniuses, is so very well-known, that it is hardly ever questioned. Therefore, we took it to be a *desideratum* of our study that it should recognize the obviousness and prominence of this fact. That is to say, whereas the gradation from the normal through the mild neurotic to the severely neurotic and further to the psychotic, makes psychosis in everyday life pretty impossible, and thereby the demarcation between severe neurosis and mild psychosis impossible, we came up with a demarcation that offered two gradations, qualitatively different, between the normal and the neurotic, and between the normal and the psychotic. As a surprising corollary, we also found the possibility of a gradation between the neurotic and the psychotic, so that we allow for a person suffering from both a neurosis and a psychosis, one mild and one severe, or one severe and one mild, or both mild or both severe.

When we started the present study we had no idea its outcome would be so ambitious as to propose to open a reform of the whole clinical classification of mental illness on the basis of a new clinical demarcation between neurosis and psychosis. We intended to follow the leads of programs offered by three independent thinkers, Ey in psychopathology, Piaget in the structure of thinking, and Popper in scientific method; we attempted to apply these ideas and others to the problem at hand, and so we went as far as our considerations have led us. We were glad to stop at the suggestion of a new demarcation; we offer no theory of infant or adolescent mental disturbance, nor any theory of etiological nature whatsoever. Likewise, we should say emphatically, the absence of any views on suggested therapy or therapeutic technique is merely an expression of total ignorance: we have nothing to say on this matter. Of course, we do hope that our work will be useful for some future efforts in these directions; we are glad not to have attempted them here.

And yet, we must confess openly, the most far reaching conclusion of this essay, our demarcation between neurosis and psychosis, is not really new; rather it integrates clinical standard practices and bits and pieces from various parts of the rather vast literature. Nonetheless, we admit, the result came to us as a surprise and as a corollary to our assault on the paradoxes of paranoia.

Finally, our thanks to Professors F. Brüll, Z. Giora and B. Scharfstein of Tel-Aviv University, to Professor C. Frankenstein of the Hebrew University, Jerusalem, and to Professor Peggy Marchi of Union College. They all read the early draft of the present study and offered both encouragement and criticisms. Mrs. Juliette Birnhack was extremely patient and helpful with the preparation of the typescript through its various stages. Tel-Aviv University Sackler Medical School and the Day Hospital and Mental Health Clinic of Kupat Holim (Ramat Chen, Tel-Aviv) helped with the expenses. Our thanks to them all.

Tel-Aviv University, Sackler Medical School YEHUDA FRIED
Tel-Aviv University and Boston University JOSEPH AGASSI

Tel-Aviv, Spring 1973

P.S. The final manuscript version was prepared for the printer after a long delay due to an unexpected (by us) lengthy and marginal criticism from various referees, including some renowned for their ability to appreciate and encourage. We did not respond very favorably to all this. Our response, we suppose, qualifies us as paranoics, and our continuing to function at large under the guise of normalcy only exemplifies the paradox of paranoia. Our ability to argue thus – to use criticism levelled against us, as an example for our views – is another symptom of our paranoia, of course. Be it as it may, the changes induced by the many comments on our manuscript were marginal, especially such as to prove our erudition regarding material that this reader or that declared relevant or unjustly dismissed and we did not. Therefore our notes and annotated bibliography contain even more irrelevancies than we originally deemed advisable. We should note, however, that whereas Freudians were hurt on Freud's behalf and others were offended on behalf of still more distant targets of our criticisms, the chief victims of our onslaught, Drs. Laing and Szasz, were particularly encouraging. Our gratitude and salutation to them.

We wish to add one more point. Every draft of every page was conceived and written jointly, except for Appendix I for which Y. F. takes major responsibility, and Appendix II for which J. A. does.

Tel-Aviv, Spring 1975

CHAPTER 1

INTRODUCTION: THE PARADOXES OF PARANOIA

Paranoia[1], *délire systématisé chronique,* is a well-known yet rare disease.[2] Almost all jokes about madmen and all stories about mad scientists are about paranoics, yet they are very rare: Kraepelin, for example, saw only 19 of them in the whole of his career.[3] The textbook tells us that the symptoms and signs of paranoia indicate the existence of a systematic localised chronic delusion, logically sustained. We shall elaborate on this definition later. Our immediate response to this definition, however, we would like to offer here: on the one hand, obviously, the definition is quite problematic; on the other hand, the definition has remained in the whole tradition unchallenged. Also, it is well-known, some paranoics are hospitalized, some not. It is customary to add the view of paranoia as covering some serious emotional trouble, but this is done as an afterthought, and without clarifying what the emotional trouble is.[4] To conclude, persecutory delusions and illusions of grandeur, are neither necessary nor sufficient for paranoia – quite contrary to commonly held views. For our part, we shall abide by all these points throughout the present essay: we shall accept the definition, its problematic character, etc. What we hope to do is only to expose the problems and try to cope with them.

That will do for a start. There are scarcely any attempts to offer extensive studies of so rare and so problematic an illness.[5] Yet this is what we venture to do here, and our reason is as follows.

Traditionally, mental illness was considered as something, if not supernatural,[6] then at least mysterious and quite extraordinary – of another dimension altogether.[7] Consequently, the mentally ill were treated as superior or as inferior beings, as somehow not needing or deserving ordinary human sympathy (which was so difficult, indeed terrifying, anyhow). Even those who, following Pinel,[8] Tuke, and Rush, treated paranoics as human, did not know how to cope with their illness: they offered much sympathy but no treatment or even diagnosis. Kraepelin was the first to be perceptive about mental illness;

he also saw it in a broader context (perhaps these two facts are strongly related). Yet the real breakthrough was Freud's, and because of the idea expressed in the very title of one of his most famous books – *Psychopathology of Everyday Life.*[9] It is therefore understandable that he studied neuroses in preference to psychoses, and hysteria more than any other neurosis; indeed, his studies of common phobias were his real clinical breakthrough, and his interpretation of ordinary dreams and of slips of the tongue and the pen of the most normal kind were his deepest and most lasting insight into psychopathology.[10] His realm was the borderline between the sane and the insane, and his crowning glory was the humanizing of not only the mentally ill, but also of mental illness itself. Yet his studies of psychosis were unsuccessful, and most hospitalized mental patients are psychotics, not neurotics; and so, strictly, Freud's study of the borderline cases is not quite relevant to mental illness proper. If we want to make full use of his discovery we may try to take cases which are indisputably psychotic, and which are as remote from the borderline as possible, in the hope of finding some human context for them as well. If the venture will be successful, then the whole range of mental disorder, the mild neurosis and the severe neurosis, the passing psychotic episode and chronic mental illnesses such as chronic schizophrenia and paranoia, all these disorders will be presentable as defects which have some representation or other in quite normal everyday situations. In brief, it is this challenge which we find intriguing.

Our choice of paranoia, then, is in part due to its being an extreme case: the case of a systematic chronic delusion, logically sustained. What makes the paranoic an extreme case however, is not his rarity (some rare cases are utterly unproblematic, like certain phobias discussed and satisfactorily analyzed by Freud) and not his intelligence (the schizophrenic and the manic-depressive may be intelligent too); paranoia is an extreme case by the very fact that paranoia is by definition a quirk of the intellectual apparatus, a logical delusion.[11] The moment we try to view any quirk as something normal in any way, then we enter a paradox[12] as a matter of course: the very notion of a normal quirk is paradoxical, because by quirk we mean a certain kind of abnormalcy. Before stating the paradoxes of paranoia[13], however, let us stress, lest the reader feel we play with intellectual teasers and sophisms, that the paradoxes are a source of both vexation and hope: the delusion is logically sustained and this may open the road to its comprehension by logical means. There is a parallelism between the intellectual disorders exhibited in the paradoxes of paranoia and the emotional disorders studied by Freud. Whereas he found an emotional context for emotional disorders and even for dreams, we seek an intellectual context for paranoia. That is to say, we hope that the very difficulty we have in distinguishing the quirk of the paranoic from that of the adolescent, of the avant-garde thinker, and of the fanatic, might provide us with a context in which to place the paranoic.[14]

This is not to say that we consider paranoia a purely intellectual disorder. Indeed later on we shall present the view that every psychopathological condition, neurotic or psychotic, mild or severe, is a disorder, both emotional and intellectual. For, every moment in life, be it normal or not, is both emotional and intellectual. Yet, generally, we shall try to view paranoia as a primarily intellectual disorder or a disorder of the thought-process,[15] and neurosis as a primarily emotional disorder.[16] There is really nothing surprising here. Folk mythology – jokes and stories – about madmen presents their disorder as intellectual. Of course, some folk stories present madness as sheer stupidity, but others do not: the punchline of the standard madman joke is well-known: I am nuts but no fool. What is the disorder, however, folk mythology does not say.

When Freud uncovered neuroses as emotional disorders, he thereby made madness more ordinary. And so, the very fact that Freud wanted to make madness a mere matter of degree gave him a bias towards studying emotional disorders, since he could show them to have much in common with the normal. And so, either he ignored the intellectual component of disorders as much as he could and hardly ever laid any stress on them as such, or he even denied their presence as sources of mental illness, namely as qualifying the patient as psychotic. But, to repeat, intellectual disorders, whether pathogenic or not, are obviously present, and obviously more pronounced in psychosis than in neurosis; particularly so in paranoia. It is thus no surprise that in addition to the fact that Freud was so much more at home with neuroses than with psychoses, he had the tendency to view psychoses too as emotional disorders, only of a severe sort.[17]

Here he obviously went off the rail.

Here we must raise the question: what is primary in psychosis: the emotional or the intellectual disorder?

It is very important to warn the reader not to confuse primary with pathogenic.[18] We find the prevalence of this confusion in the literature something which requires a radical treatment.[19] We use the terms "primary" and "secondary" not as used by Freud in his early *Interpretation of Dreams* and posthumous *Outline of Psychoanalysis*.[20] Rather we use them as in the medical literature, or as in Freud's earliest – and medical[21] – publications. There are in the neurophysiological literature well-known principles, Jackson's principles by name,[22] which say, whenever a stage in the neurophysiological hierarchy is damaged, all higher stages are put out of action, and their effects are not seen any more, whereas actions of lower stages which are inhibited by higher stages and which can operate as functional and integrative, show up again.[23] The absence of the higher stages is called primary, the appearance of the lower stages released by that absence is called secondary. Obviously, neither is pathogenic, but both are caused by some (unspecified) damage (which may be caused by diverse causes,[24] but which is not discussed in the principles).[25]

To return to our question, what is primary in psychosis, thought impairment or emotional impairment? In our opinion, as in the views of Bleuler and others, for schizophrenia, the primary impairment, and that which qualifies it as psychosis, is in the thought process. We feel the same about paranoia. We therefore begin by accepting Ey's serious recommendation to study the intellectual apparatus of the patient.[26] With this, we think we can now start with the presentation of the paradox or paradoxes of paranoia.[27]

We present the paradox or paradoxes of paranoia in four statements, which may draw our attention to different aspects of paranoia.[28] The four paradoxes of the normal structure of the paranoic mind have in common the fact that we perceive in it qualities which are condemned therein, but lauded elsewhere. Here are the – normally commendable – qualities which are condemned in every manifestation they have in each case study of paranoia: (1) The paranoic is logical. Indeed, he is strikingly meticulously logical. (2) The paranoic perceives well and correctly, perhaps even accurately. Indeed, we have well known cases where paranoics' perceptions, surprisingly enough even of their own conditions, are of unusual sharpness and acuity. (3) At times fundamental assumptions of the paranoic are no worse than that of the alternative – those which are accepted by his society (the popularity amongst paranoics of the Christ[29] complex[30] is no accident). Of course we view as a mere curiosity any case of paranoic ideas which later on gain popularity; and rightly so. Yet there is an injustice here. (4) At times the fundamental assumptions of the paranoic are integrative principles which make his image of the world an integrated one, better than the average. Now, we do not have to be metaphysicians to appreciate the integration of a person's view, and yet we are usually amply justified in rejecting a paranoic's integrative principle. Oddly, perhaps, it is not at all easy to say why.

That there are other aspects of paranoia, in which the pathology is obvious, cannot be doubted for even one moment. These are perhaps unrelated to the above, for example depression, hesitation, aggression, etc. (nevertheless, we shall relate them to the paradoxes – as secondary – in the clinical chapter below). Or they may be contrary to the paradoxes, for example they may be highly dissociative (these too will be related there in the same fashion). It is, indeed, because of the variety of symptoms and signs noted here, that the paradoxes, though well known to some writers, have not been presented in the literature so poignantly, and collectively, as in the previous parapgraph.[31] Moreover, of those writers of whom it may be said that they took these paradoxes seriously, all except Ey have resolved them by denying in effect that the mentally ill suffer from any intellectual defects in the general sense – often quite inconsistently. We have in mind in particular two very important authors, R. D. Laing[32] and T. S. Szasz.[33] Although we fully accept their moral commitment to the equality of the mentally ill as a member of the community, we do not think that either of them has offered a theory which resolves satisfactorily the paradoxes of paranoia. Indeed, we consider the paradoxes of paranoia paradoxical precisely because on the one hand we recognize the formal defects in the general structure of the intellectual apparatus of the paranoic, and on the other hand we consider the paranoic's mental apparatus superior, at times superior in every one of the four respects which we have just enumerated. And so, to begin with, we bluntly class paranoia as mental illness and mental illness as illness proper;[34] this quite unlike Laing and unlike Szasz. We shall endeavour to argue that, paradoxically, the paranoic's weakness lies in this very superiority. It is no accident that paranoics are so often very logical, so much so that many take this fact for granted as part and parcel of the definition of paranoia. This difficulty is indeed the crux of the present study.

Perhaps we should say this from the onset. Both Laing and Szasz (also M. Foucault), each in his own way, take the paradoxes seriously enough to deny the very existence of the paradoxical entity, mental illness. They thus force us to see the person we label as mentally ill as a merely social deviant – of one sort or another. We cannot go into this in sufficient detail here: it requires a whole chapter. And so, Chapter 3 will deal with it. All we can say here is that we disagree: were paranoia no psychological problem at all, we would not be writing this study on it.

A few more points before we conclude this Introduction. As the title of this study suggests, we are treating paranoia exclusively, and as this introduction suggests, only an aspect of it, namely only the paradoxes. However, clearly, the reader expects more and the authors aim at more. Also, the learned reader may be puzzled by the fact that we include R. D. Laing and Thomas S. Szasz in the list of commentators on the paradoxes – we shall indeed discuss their views extensively later on – whereas they speak in general of mental illness, and in particular of schizophrenia (Laing), of hysteria (Szasz), not of paranoia.[35] The reader who is a practitioner may have the same worry not so much *à propos* of Laing and Szasz as *à propos* of his empirical experience: he may wonder whether we intend to center strictly on paranoia, *paranoia vera* so called, and arbitrarily exclude its ramifications elsewhere, or whether we intend to include the common and multiple forms of paranoid phenomena, delusions, persecution, illusions of grandeur, etc. as they may occur in schizophrenia, depression, senility, toxic psychosis, paranoid personality disorder, and even in mere adolescent crises. If we take only *paranoia vera,* our study may be too narrow. For, is not *paranoia vera* so very rare, and thus better taken as a mere curio? Yet, if we take paranoia in the broad sense, we may be trying to do too much. For, is paranoia not somehow involved in all mental aberrations?

Admittedly, *paranoia vera* is a rare disease. But it is not only that. It is also an extreme case in psychopathology: extreme and problematic. Let us elaborate a bit on this point before returning to our discussion. Paranoia is extreme and problematic since even its being a disease is elusive: in paranoia the diagnostician can hardly pinpoint where the disease lies. Two circumstances account for this difficulty. First, the patient is hardly a patient in the usual sense of the term: generally, a patient is one who comes to the physician and complains, whereas in paranoia the patient normally does not come to the doctor at all, since he does not consider himself ill in the first place.[36] There is no "presenting symptom" in the medical sense, since the patient is not complaining of anything "medically": he might, when pressed, complain of being persecuted, yes; but since when is persecution a medical complaint? And when psychiatrists examine psychiatrically the paranoic's complaint, then the second circumstance pops up, which make the very illness that is paranoia so elusive: apart from the delusional[37] system we find no psychiatric disturbances. Indeed, according to E. Kraepelin's famous definition of paranoia of 1893, the patient's logic, affect, willing, and action, are all well preserved.[38] This, of course, is the very root of the paradoxes of paranoia, as already discussed, and as will be discussed again repeatedly later on. All this, to repeat, may raise the question, why bother with so rare and difficult and problematic a disease?

For our part, we think all extreme cases are not only challenging, but also inviting subjects for fruitful studies of much more general applicability than initially seems plausible. Thus, we believe that the study of paranoia – precisely because it is the extreme case – may provide us with a clue to a theory of demarcation of mental disease; of mental health from mental disease.[39] Likewise we hope to demarcate neurosis from psychosis. We shall consider paranoia as a highly integrative state, but paradoxically still dissociative (we shall offer a more precise study of the nature of the dissociation; and indeed we shall suggest it as the very first line for the demarcation of psychosis).

Other forms of paranoid delusions (in schizophrenia, depression, senility, toxic psychosis, paranoid personality disorder and adolescent crisis) will be later easily defined as more advanced states of dissociation, yet those clinging to an otherwise (underlying) integrative principle. In other words we shall offer the view of paranoia as rare only in its purity, not rare otherwise, i.e. having much in common with all other psychoses and even with neuroses.

To conclude this introduction, we hope that the reader will agree to share with us the readiness to take the paradoxes of paranoia as seriously as possible, regardless of the rarity of the disease and independently of the generality or otherwise of the solution we may come up with. We confess this is the way we began the present study; and whether the generality of our conclusions is justified or not, we confess it surprised us as it unfolded itself to us in the process of the study here reported.

CHAPTER 2

PSYCHOLOGICAL BACKGROUND

The phenomenon of delusion is too obvious to have ever been discovered.[1] It was usually ascribed to either supernatural causes or to derangement. Before the French Revolution[2] no mental illness was taken seriously[3], as the first to treat mental illness as an illness, respecting its bearer as a human being, were the clinicians who attempted to implement the philosophy of the Enlightenment.[4] Only about a century later, Kraepelin took the phenomena of mental illness, including the varieties of delusion, seriously enough to describe and to diagnose some of them as paranoia.[5] Like everyone else before Freud, he too was convinced that brain damage (or organic cause) was the basis of every mental illness. (This is the meaning of the 19th century materialistic slogan, "No psychosis without neurosis", where "psychosis" means mental damage to non physical entities and "neurosis" means physical damage to such physical entities as nerves and nerve centres[6].) Nevertheless, Kraepelin miraculously noticed the paradoxically logical strength of the deluded mind. He tried to explain this by the observation that the deranged mind possesses some defective faculties but some undamaged ones, perhaps with the improvement of the latter to compensate for loss in the former. He could not relate the defective qualities to any brain damage, but he was able to pinpoint the defects; and we are not yet able to improve on him in this respect. What he said was very briefly as follows: The paranoic's logic is perfect; his premises are false. To use his own oft quoted words, paranoia is "the insiduous development of a permanent unshakeable delusional system from inner causes in which clarity and order of thinking, willing, and action, are completely preserved."[7]

This marvelous sentence is highly problematic of course. Indeed, Kraepelin himself is aware that it is too broad: after all we do not call paranoic all racists, xenophobes, crackpots, etc. And as if to exclude these he adds a qualifying sentence to the above quoted sentence: "It [the delusion] effects a deep seated change of the total outlook on life, and a derangement of standpoint towards the surrounding world".[8] In other words, we take this as a qualification of the definition, and so, obviously, as part of the definition. Indeed, in Kraepelin's definition of paranoia[9] we find two sentences: Paranoia is a disease which consists of (1) an "insidious development of a permanent unshakeable delusional system from inner causes, in which clarity and order of thinking, willing, and action are completely preserved." (2) "It [the delusion] effects a deep seated change of the total outlook on life, and a derangement of standpoint towards the surrounding world." Now, it is clear that the first sentence gives the first part of the definition – *per genus* – while the second one gives the specificity of the condition – *et differentia specificum* – so as to differentiate delusional systems of any other kind, from that of the one which "effects a deep seated change of the total outlook on life", of the paranoic. More specifically, "It" should be replaced by "such that it" and the two sentences merge into one. And, to be pedantic, we should notice clearly the different implications of the two alternatives, first, "Paranoia is delusion. It causes derangement", and second: "Paranoia is delusion which causes derangement." The first alternative forces us to view all crackpots as paranoics doomed to derangement, the second alternative raises the question which delusion causes derangement, which not, and why.

To return to Kraepelin, his problem began when he noticed the paradoxical fact that paranoics have powerful logic on their side. Before him people could view paranoics as deranged, deluded, illogical, etc. (See Ibsen's *Peer Gynt).* Only when he noticed the logicality of the paranoic he asked, why do we consider him as defective at all? And he said, the logic of the paranoic only enhances his error. But then, of course, the ordinary logic of an ordinary crackpot must lead to the same derangement, perhaps a bit slower. Indeed, all of Kraepelin's pronouncements clearly indicate that he had to conclude that all delusion, even that of the crackpot, must finally lead to derangement; yet he wanted to resist the conclusion, being fully aware that not all crackpots are deranged. Without going into too many details, let us simply quote other passages from Kraepelin which exhibit the same seeming awareness, namely that the delusion is a disease, consisting of a thought built on false premises, falsely presumed due to the disease,[10] "every delusional idea is a representation falsified by the disease,"[11] and "the delusional ideas are mistakes aroused pathologically, which are inaccessible to rectification by rational means".[12] Let us not go further here into the source of Kraepelin's systematic ambiguity.[13] We shall discuss in the next chapter the intellectual background to Kraepelin's difficulty and show how hard it was to avoid it.[14] For the time being we should move on to Freud's contribution.

When we come to discuss Freud, we must mention what we consider his greatest insight[15] into psychopathology in general, or rather into all genesis of psychopathology; an insight which makes sense, however partial, of neurosis and psychosis alike. It is the insight which we make extensive use of throughout the present study. We are referring to Freud's idea that every neurosis and every psychosis is a defective mode of adjustment, which somehow cannot be easily improved upon and so acts as a trap, which is self-reinforcing and so acts as the illness and the treatment at one and the same time. To be specific, a defence mechanism is a mode of treatment, a defence proper, though a defective one. Yet, we view it as pathological because it is quite unnecessary from a purely objective viewpoint, but itself creates the necessity. Let us take a very simple example. Shyness, in all its neurotic manifestations from stuttering to hysteria, is a well known phenomenon, shared to this or that extent by all, especially in childhood. It is, need one say, closing oneself against others, seemingly out of their aggression. Now this feeling is obviously self reinforcing simply because it prevents one from getting used to others, since not meeting people one is not immune to their small aggressions and so is more in need of shyness which is thus both the ill and the defence against the ill.

Now, this view of the role shared by neurosis and psychosis immediately raises the question, what is the difference between neurosis and psychosis? The old meanings of the words were lost with Freud's rejection of the view of all mental illness as caused by organic defects.[16] In addition, there is no rule in Freud's theory of psychopathology for saying that a patient's attitude to this phobia, for example, is in one case neurotic, and in another psychotic.[17] Nor is there any reason to put one form of narcissism, which in his view is schizophrenic,[18] as psychotic, and another, which in his opinion is hysterical, as neurotic.[19] Nor is there a theoretical basis in Freud's works for calling hysteria a neurosis[20] and paranoia a psychosis.[21] The nosological distinction between the neurotic and the psychotic is in Freud's works, and up to this day, both a matter of clinical picture, and a matter of etiology and pathogenesis.[22] This, however, only raises the question, must these two – clinical picture and psychopathology (be it etiology or pathogenesis) – always go together?[23] Could we, in other words, stick to a diagnosis of a symptom as neurotic even after we declare the patient a psychotic? And could we do it the other way round? The psychiatric consensus today is that this is impossible, that diagnosis and etiology must go together, that a psychotic patient can have only psychotic symptoms and a neurotic patient can have only neurotic symptoms, that when we find a conflict between diagnosis and pathology, for example when we hear of a psychotic patient with neurotic symptoms, there must be a mistake somewhere. The question is, why? We could well imagine a psychotic breakdown occuring in a neurotic patient, or the other way around. (And indeed, in our own view this is quite common.) Thus far, in practice, all those who had breakdowns or psychotic episodes and were clinically diagnosed as psychotic, are declared as suffering from psychoses. Yet there is no theoretical justification for this rule, much less an empirical foundation (since, as we already stated, the empirical findings are summarily rejected when they fail to conform to the rule[24]).

Freud not only failed to demarcate neurosis from psychosis, he also failed to explain the rule of correlation just mentioned, between diagnosis and etiology or nosology. Indeed, inasmuch as he offered treatment for psychotics it was essentially the same as the treatment he offered for hysteria – a classic neurosis. The consensus is now increasingly towards the view that he offered no treatment for psychosis[25] (some even say he refused to treat psychotics).[26] To this the following proviso is often added. Anyone who succeeds in applying Freudian treatment to psychotics is praiseworthy, either because this may be possible after all, or because when a schizophrenic is psychoanalytically treated with a considerable measure of success (not to say cured) he can be viewed as a neurotic who only seems to be a psychotic. For our own part, we find all this painfully apologetic. We think one fact is obvious and needs stressing: Freud did view psychotics as if they were hardly different from neurotics, and he was mistaken here. Now we find no need to defend this great man's reputation – certainly not by hiding his mistakes and shortcomings, nor by ignoring his less successful valiant attempts to help his patients.

Freud may have been right; yet we declare – in accord with professional public opinion – that psychosis, especially paranoia, is a disease of a different order or dimension from neurosis. Some clinicians, more so some vulgarizers, demarcate neurosis from psychosis by saying that the later but not the former requires hospitalization. And, indeed, this is statistically the case. But it is not always so, and even the fact that most patients in mental homes are psychotic may be a transitory situation. Rather, we claim, in psychosis the patient's outlook is more radically affected than in neurosis; that the patient's defective outlook is what makes him a psychotic. We follow Ey in saying that the psychotic disturbance is in the very structure of the psychotic's mode of thinking or of his mental apparatus.[27] Whether one accepts this or not, one usually finds it hard to describe exactly how the psychotic illness goes deeper than the neurotic.[28] We have already suggested that the causes of this difficulty are the paradoxes which stress the intelligence and strength of the paranoic's thinking (and to a lesser extent, we shall later argue, even the schizophrenic's thinking). We should only add that the high correlation between diagnosis and pathology of thought in psychosis is cogently explained by the view of psychosis as primarily an impairment of the thinking apparatus; particularly so, because, as we shall illustrate later on, in our view one should consider as a mere approximation all sharp separation between the intellectual and the emotional.

This ends our discussion of Freud. (See, however, Appendix II to Chapter 8 below.)

We could stop here on the pretext that there is no authority on paranoia in a class with Kraepelin and Freud. Yet we would do an injustice in this way to more modern writers who did so much with the paradoxes of paranoia. Let us at least mention Thomas S. Szasz, and Ronald D. Laing.[29] Before that we feel we should mention Melanie Klein's attempt to view psychosis as a common malady, akin to Freud's view of neurosis.

Melanie Klein's theory is a variant of Freud's: she uses as given building blocks Freudian elements, though she disagrees with him on quite a few issues. What concerns us here is her view of psychosis as common to all newborn infants and thus common to all adults in varying degrees. Let us only remind our reader that Melanie Klein's work[30] is notoriously obscure and difficult and subject to much interpretative study; we shall here rely on the expositions of Hanna Segal[31] and J. O. Wisdom.[32]

Klein observes the newborn infant on the assumption that he possesses as much adult abilities as possible. The infant does not, however, possess knowledge – not even of the existence of his mother. He begins with positive and negative feelings of all sorts, with an ego and the ability to project and to internalize or introject. The result is an extreme polarization of the universe into the good and the bad,[33] a condition she calls paranoid-schizoid. It might be called more accurately friendly-hostile-schizoid. But as the hostile is identified with the persecutionist and the persecutionist with the paranoid, the hostile is labelled paranoid. And the friendly is totally omitted as rather unproblematic in spite of the well known fact that paranoia may be expressed as a picture of all-pervasive and overflowing goodness and kindness – of which the paranoic may view himself donor or recipient or both. (The television program 'The Invaders' is paranoic persecutionist; its mirror image, 'Mission Impossible' is equally paranoic yet its task force is utterly benevolent.) And so, the title schizoid-paranoid, inadequate as it certainly is, brings home the claim that we are all psychotic even before we obtain knowledge of the existence of others – our mothers in particular. The appearance of the knowledge of the existence of our mothers, says Klein, leads us to a real crisis: despair at our helplessness, fear of desertion, and later on jealousy. The bitter disappointment, however, is not without its benefit: the infant learns to distinguish truth from falsity in the very process of reparation, i.e. of the emergence from his depression.

Here, for once, Klein's negative approach pays a dividend. Her theory begins with attitudes symmetrical with respect to good and evil. The symmetry is somewhere broken, and rather unsatisfactory. But in one place attitudes obviously cannot remain symmetrical – not regarding truth and falsity. This is so because the idea of falsity in infants is identical – we all assume this – with disappointed expectations. For when a good expectation is disappointed it must bring about hostility, whereas when an evil expectation is disappointed it need not be replaced with friendliness: the hostility may be perpetrated by projecting anxiety and even if hostility vanishes it need not arouse friendliness (though great relief does naturally arouse friendliness). Admittedly, when expectation is fulfilled in a newborn infant the accompanying feeling is allegedly sustained; this only shows that whereas in truth the symmetry between good and evil is sustained, in disappointment the symmetry is shattered. And so, the infant is pushed more towards unfriendliness, by this theory, than towards friendliness, even without any trauma or other drastic evil. Of course, conceivably mothers may see to it that a child is amply soothed and compensated so as to maintain the balance between good and evil. The point is, if she does not actively try to pacify her baby he already tends to be psychotic.

Freud's stage of ego formation is pushed by Klein back to the time of birth; his Oedipal stage from the genital to the mid-oral phase; and his formation of the super-ego stage is pushed by her even further back from the Oedipal stage to the early depressive stage: when a few-days-old child learns of his mother's existence and he gets depressed and jealous, he also learns to develop love and responsibility. This development happens when he first repairs his depression. Alternatively, he may shift from depression to mania, where the relation of mania to depression is an imitation of the original relation between good and evil, namely the relation of polarization which constitutes the initial schizoid split. Hence, the manic depressive state is but a variant of schizophrenia. Hence, and this is important, psychotic cases may be viewed here as mere Freudian regressions.[34] We shall later sustain these conclusions, even though we shall not consider child psychology or any theory of the causes of psychoses, Kleinian or otherwise.

The question may be asked, what now of the distinction between neurosis and psychosis? To answer this in detail we would have to plough through the Kleinian literature.[35] Suffice it to say here, then, that at least at first blush Klein has turned Freud's theory upside down: whereas he sees only neuroses everywhere she sees only psychoses everywhere. Whereas Freud assumes the newly born to be perfectly healthy though soon to be damaged to this or that extent, wounded or scarred as the case may be, Melanie Klein sees the newly born as utterly psychotic though with the ability to adjust and cure oneself and become normal to this or that extent. It must therefore be hardly surprising that Melanie Klein undertook the treatment not only of children – in the course of which she produced her ideas – but also of adult psychotics, and with some reported measure of success. For our part we can hardly assess the success, not knowing how she, or her champions, distinguished psychotics from neurotics.

We shall later endorse Klein's theory of psychosis as a regression but only after distinguishing clearly neurosis from psychosis. We shall also accept the corollary that there are gradations between normalcy and neurosis. But we shall discuss in this study neither infant-psychology, normal or abnormal, nor etiology, nor treatment.

There is hardly any need to draw attention to Klein's influence on R. D. Laing or of Laing on Szasz. We shall discuss, however briefly, these two authors' views before bringing this chapter to a close.

What Szasz[36] did was to close the debate on the gradation between normalcy and mental illness by denying that mental illness at all exists, or, if you will, by saying that we are all subject to the same strains and stresses, wear and tear, as those who declare themselves mentally ill. If this position sounds extreme, it is a mild one as compared with that of R. D. Laing[37] who, in all seriousness and with great ingenuity, defends the famous paradox that only the insane are sane. But let us take Szasz first, though chronologically he comes second.

It is difficult to say whether Szasz flatters the madman or cows him when he declares him normal. In declaring him normal he does not, of course, deny that he needs help; he declares him normal only in the sense of ascribing to him the full responsibility of a 'normal' citizen, both the rights and the burdens of responsibility. From this to the paradoxes of paranoia is a stone's throw; and Szasz knows this only too well though he does not concern himself with these to any extent, since he speaks of all mental illnesses equally, psychotic or neurotic.[38] Moreover, making it his business to criticize Freud, he naturally ends up discussing in a book on (alleged) mental illness almost exclusively the case of hysteria.

Whereas for Szasz the paranoic is normal, for Laing the so-called normal is the one who is really sick. To prove this Laing uses the armory of existentialist philosophy, especially the concept of alienation. This may sound like forceful argumentation in existentialist circles, but it is neither here nor there for non-existentialists who happen to think we are all alienated, normal and abnormal alike.[39] Laing also uses straightforward arguments. He puts forward explicitly as arguments for his view certain consequences of the paradoxes of paranoia.[40] This, incidentally, is paranoic. It was declared to be so with a very unprofessional glee and *Schadenfreude* by quite a few professional reviewers of his works. The very polarization of Laing's readership into fans and foes, incidentally, is paranoid-schizoid, of course. We really think that at least the psychiatric profession may be expected to handle matters like that in a more levelheaded manner.

Laing presents a paranoid-schizoid view of our society. He brilliantly describes society as polarized into original spirits and conformists, with the latter, of course, in the majority. The conformists are the hateful living-dead who persecute the beloved original spirits.[41] It is somewhat disappointing that Laing should fuse paranoia and schizophrenia so systematically; it is even more disappointing that he should consider all original people as anticonformists, all anticonformists as original, all originals as schizophrenics, etc. etc. Yet, to apply the paradoxes of paranoia to Laing, certainly his view is an excellently integrated view,[42] not always inferior to accepted views, and all too often superior to the views of his reviewers; it is a view supported by sharp logic and by profound perceptions. Thus, all our paradoxes of paranoia apply to Laing's view which is rejected as crazy in spite of all the great assets it obviously possesses.

This is not to say that Laing's views are merely views of a paranoic: the paradoxes are so shocking just because (as Bertrand Russell has noticed; see note 37 to Chapter 1) they apply to sane and insane alike, and paradoxical views have often been rejected because of their assets not only in psychology but even in mathematics and physics. Those who denounce Laing's views on the strength of paradoxicality have failed to see the force of the paradoxes of paranoia in general, quite independently of Laing's own views. For our part, though we do not endorse Laing's view, much as we value it, at the very least we agree with him that fear to see the strength of the paradoxes of paranoia is, like any other intellectual fear, all too human. Except that this fear is harmful in that it puts the patient in painful isolation. And this brings us to the point which Laing recognizes but plays down, yet which Szasz makes prominent: the so-called mentally ill pretends to have lost his sense of responsibility merely in order to compel society to take notice of him and his sufferings. This Laing cannot accept, as he blames society for something much deeper than neglect:[43] he blames society for the patient's very need for help, since the need stems from his suffering from the excessive demands to conform which society places at his door.[44]

Laing is a protector against the injustice rendered to man by society. Szasz is much more realistic. If he complains, he complains against the defects in the law which permit involuntary certification and hospitalization,[45] or which reward criminals for their abdication of their reason and sense of responsibility. But he is much more concerned with the patient as a patient.

For Szasz the patient's need for help is what makes him a patient.[46] He puts much emphasis not only on the need for help, but also on its very expression: we cannot legitimately offer help without being asked. Szasz differentiates between the neurotic and the psychotic, by differentiating their modes of plea for help: the neurotic's plea is in common language, yet for uncommon causes, whereas the psychotic pleas are for very common causes, yet in a peculiar and private language of symbolic actions. In other words, the neurotic pleads in normal language for abnormal causes, and the psychotic goes the other way round. We accept this categorization, but consider it secondary:[47] we shall later venture to explain this by reference to primary impairments. Meanwhile let us note that Szasz's theory explains the high intelligence-level required from the psychotic. It also explains the depth of his trouble as a very basic need, a very normal common one, strongly felt since seldom gratified since seldom comprehensively articulated. It also explains the use of quaint language as rooted in the psychotic's own fear of comprehending his own signal. For these facts which Szasz's theory explains we have a great deal of sympathy, and we shall venture to offer a theory which does justice to the same facts, yet which locates the impairment of psychosis primarily in the mental apparatus and not necessarily at all its means of articulation.[48]

Szasz's major thesis is that there is no defect in the mentally ill which permits us to ignore his status as a responsible citizen. Now the very demarcation of psychotic and neurotic in terms of their expression is, of course, an admission of their peculiarity: not only do they suffer, but also they ask for help in one strange way or another. Is this a symptom of a defect or not? If it is, is the defect one which impairs the personality and responsibility of the patient? Szasz is hardly concerned with this. He presses his desire to help and his willingness to treat the patient as an equal, and he leaves his argument only half-articulated. Szasz does not care much for analyzing the patient's mental apparatus: his business is primarily to re-establish the patient as a full member of society, and secondly to alleviate his pain.

This is a point not of criticism, but first of agreement on responsibility being the major point and, second, of recognition of and respect for Szasz's awareness of his own limitations. He stresses a few times that he is "emphasizing the urgent need to clearly specify norms and values first, and techniques of behaviour second".[49] It is because we accept this maxim, as well as the same norms as he does, that we go on to discuss techniques of behaviour. We agree that mental patients should be rehabilitated, but, for our part, we frankly doubt that all psychotics can be rehabilitated, though perhaps with Szasz's help we may arrive at a time when no patient will be neglected and left to deteriorate that far.[50] Also, contrary to Szasz, we think that the paranoic's intellectual apparatus has to be better understood in order to achieve more effective treatment. We confess we find in Szasz's disregard for the psychotic's, especially the paranoic's, crazy ideas, an old fashioned ring. We shall discuss and criticize the old fashioned views of madness in the next chapter.

CHAPTER 3

SOCIOLOGICAL BACKGROUND

What tempts us to present the sociological background to our study is the fact that for so many paranoic beliefs these exist communities of normal people holding these very beliefs. This is so not only in the weak sense that most members of a given community may believe in magic, ghosts, goblins, cargo from Heaven, saviours etc. It is so in the additional strong sense that a view which is considered paranoic in one society may be institutionalized in another society and become obligatory there: to deny it there would be criminal or plain crazy.

This fact is recognized, for example, in English penal law, where killing due to a mistake is excusable if the mistake is shared by the killer's peers and community, but not otherwise. This leads to fine points of law, raised by borderline cases, such as the well known case of the members of the sect of Jehovah's Witnesses who refused medication for their sick and dying child and who were brought to trial for criminal neglect. We need not go into that case here beyond noticing that even the most radical proposal to alter the notorious M'Naghten[1] Rules,[2] does not go so far as to declare any Jehovah's Witness insane. And yet, under some conditions we will unhesitatingly diagnose such a person paranoic. Among these conditions there will be the one specifying that he is not a member of such a community, yet shares their view, and even holds fast to it.[3] His sticking to it in the face of an offspring's death may at times be evidence that he suffers from paranoia, at times that he is a fanatic.

We wish to stress the empirical fact: a person may be classed as paranoic for holding a cranky view, but not if he belongs to a community in which that view is institutionalized.[4] We do not think that anyone will contest this statement of empirical fact, since quite a few writers, both literary and psychiatric, have ventured to make sense of it. Yet, we think, no explanation of it thus far is even remotely adequate.[5] We shall offer an adequate one later on.[6]

The (admittedly too glib) explanation most literary commentators offer regularly, and even psychiatric commentators offer at times, is that the madman who conforms to the mad views institutionalized in his society is no less mad because of this, yet he is protected by a law of large numbers: just as we simply cannot arrest all violators of the law if there are too many of them, so we cannot commit all madman if there are too many of them. Hence, though technically we cannot incarcerate all Jehovah's Witnesses, it is still the fact that they are mad and could use some medical care and treatment. Now, this may be a cynical view about conformism or a spirited protest (Laing-wise) against conformism. More seriously, it may be a view of how sociologists approach societies: they may be people who accept norms of societies as they find them, and so take them, *inter alia,* as local standards of sanity, regardless of the mental sufferings involved. This, however, is not the case for sociologists, though it is for social anthropologists. Unlike social anthropologists, who zealously defend all primitive tribes' rights to perform their magic rites, sociologists (except for sociologists of knowledge) are not usually patient with wide-spread prejudices, whether against medicine or against foreigners. Morris Ginsberg is typical in this respect: he has attempted to study prejudice in an effort to combat it. The strength of a genuine and troublesome prejudice, he says, is that it offers a genuine integrative principle which turns the view its holder has of the world into an integrated unit.[7] That is to say, the holder of a prejudice sustains it systematically and logically. For example, the holder of a prejudice against doctors sees not only doctors in a different way from us; he also sees differently facts which we may wish to use as an argument against him, including facts about health and disease, so that he sees the same facts as innocuous or even as supportive of his view. What then can be done, asks Ginsberg? Nothing short of a total and systematic attack, an attack on all parts of the integrated view at one and the same time so that it all collapses at once with nothing remaining for a reconstruction.

Now, how universal is this case? We all know of cases in which we argue against friends: by raising arguments which they were previously ignorant of we bring them to alter their minds. These friends are surely not prejudiced. At times we meet people, be they friends or colleagues who have heard of our arguments yet find them innocuous, just as the prejudiced do. Can we therefore brand them as prejudiced? Not always. If we can invent a new argument which they have never heard of, then we can have a test case of sorts. Arthur Koestler and Michael Polanyi mention the fact that a Marxist, or a Freudian, if he is orthodox enough, will have a reply to a new argument ready even before the new argument is fully stated. Can we nevertheless alter the view of a person who normally practises such techniques? Ginsberg is optimistic; others are less optimistic. In particular, Popper has claimed that there is a technique for turning any new criticism into a supporting argument, and he has called this technique reinforced dogmatism.[8] Ginsberg thinks the whole system, including its reinforcement, can be attacked as a unit and a better replacement offered and at times accepted. Prejudice is, indeed, a systematic error logically sustained, but it can be systematically attacked and replaced with a systematically better view.

The obvious question, then, is, can the same be applied to paranoia?[9] Ginsberg's definition of prejudice is almost identical with the already mentioned Kraepelin's definition of paranoia.[10] This, indeed, is no accident. To explain it we need to go to the roots of modern sociology in 18th century psychology: we should note in this way that the terrible identification of paranoia and prejudice with deviation, this ultraconservative Procrustean bed (which so irritates Laing), came as a slight alteration, even as the outcome of an effort at a slight improvement, of a very enlightened progressive view!

The view in question, the Enlightenment theory of Man, is a very influential one, not only in that almost all contemporary theories of Man are variants of it, but also in the sense that it repeatedly re-emerges in part, or even as a whole. Thought it is an amazingly optimistic philosophy, it is reflected even in writings of pessimists like Freud.[11] According to the Enlightenment, man is a quite unproblematic creature. He has few basic needs, he is naturally friendly, and he has a strong urge to acquire knowledge. It is hard to believe that such a fundamentally simplicist view could ever be entertained by the learned world. But this simplicist view is, of necessity, rather remote from experience; its importance is in the function it serves as a metaphysical integrative principle. On a more empirical level the picture of man is vitiated precisely by the defects which we observe empirically. The way the normal defects of man are introduced is the way which, incidentally, Kraepelin ascribes to the paranoic and Ginsberg to the prejudiced. Man may make mistakes, and even fall in love with them. The mistakes then vitiate by becoming integrative principles which are logically sustained, and whose distortive effects cease to remain localized, and spread with almost every act of logical reasoning. When the mistakes have thus become global, they are almost utterly impossible to shake off.

It is hard to consider the role which error played in the philosophy of the Enlightenment until one realizes that it is supposed to turn the admirable creature of nature which the Enlightenment optimistically considered Man to be, into that corrupt creature of society which we encounter daily. To this end, error was made a very powerful evil and a very tempting and persuasive evil at that; an evil malady which is as hard to get rid of as paranoia. The more we see Man as wholesome and good, the more our ordinary image of ordinary people is split and polarized into the wholesome Man and his terrible error. Were this view of evil not so popular, we would easily be able to diagnose its adherents as paranoid or as paranoid-schizoid. The very polarization of Man into a primary good plus a secondary accretion of evil, this very crazy idea, is what made it possible for the Age of Reason to hope for perfection,[12] and indeed to hope that Man will achieve perfection soon by the very small step of willingly giving up all errors. Admittedly, errors are hard to get rid of. Yet, once one realizes, that by one step one may achieve so much, the effort would become very worthwhile. (This is the global attack on prejudice as recommended both by Sir Francis Bacon in the early seventeenth century and quite recently by Ginsberg.[13])

It is not clear to this very day, whether the giving up of one's error is a mere prosaic act of courage, or a heroic act of conversion. There is a deep-seated ambiguity here, and we can trace much of the periods and later mental unrest to this ambiguity – but not in the present study.

A good example of the Enlightenment theory of mental illness as systematic error, can be found in Dr. Samuel Johnson's *Rasselas*.[14] In his travels with the Ethiopian princess, Rasselas chances upon an old astronomer who thinks he can command clouds to rain. He had tested his view, and, by accident, confirmed it. He decided to test it no further, since by commanding a cloud to rain in one place, he deprived another place of its allotted share of rain. He was willing to divulge his technique to Rasselas, but only as a secret and as one not to be used. Later on Rasselas explains the astronomer's misfortunes to the entourage, and draws the moral of the dangers of phantasies. Dr. Johnson himself, incidentally, being a man of imagination, was constantly in fear of losing his mind. (He was, also, both a very sick man and a hypochondriac. Yet he lived to a relatively ripe old age and retained his sanity to the last.)

For the Enlightenment, the polarization of Man into a basic good and a secondary evil could only be maintained by another strong polarization, between science and error. (We shall come to this later in more detail.) For, Man's quest for knowledge is primary, and his error secondary. What one has to replace one's errors with, then, is the recognition of Man's ignorance and the refusal to remedy it by anything short of scientific knowledge proper. Science thus became the body of theory which is absolutely free of all possible error, namely the body of (empirically or mathematically) demonstrated theories. Hence every enlightened person must endorse all and only the scientific theories available in his day. The terrible conformism[15] that the Enlightenment thus required of its members was bearable both because it was in the name of science and because it was imposed only in the very narrow domain where science was supposed to have given its final verdict. Once it is shifted from the Enlightenment background to that of Romanticism or to any other background, it becomes an intolerable requirement, since it becomes a requirement to conform strictly to rather arbitrary or accidental norms. Historically, the Enlightenment was replaced for many thinkers by the Romantic philosophy; this philosophy permitted some, though only very few, individuals to be deviants and get away with it. These are the people whom the Romantics recognize as heroes, as the only true individuals. These individuals could change the norms, or rather improve them, and thus make them a little less arbitrary, and so a little less intolerable.

Can Romanticism identify the madman in any other way than by his being a deviant? No. Can the hero be other than a madman for Romanticism? No. Romanticism accepted from the Enlightenment for no good reason the view that all deviants are madmen and all madmen are deviants, except that for the Romantic the deviant is the nonconformist instead of the non-scientist. How, then, does Romanticism single out the hero from the rest of the deviants or madmen? In only one way: by recognizing the fact, if and when it is a fact, but not before, that Society has accepted his deviant creations – artistic, political, religious – as the new norms.[16] The hero, therefore, must begin as a madman and end – usually posthumously – as a king.[17] The pressure on anyone to conform is then justified by success, if successful, and it is likewise justified by failure if failed: it is therefore always justified.[18] Yet, for it to be always justified it must first polarize people into two groups: the ordinary people who are absolute conformists and the non-conformists who are absolute madmen. Romanticism thus first polarizes humanity into conformists and non-conformists, and then it polarizes non-conformists into heroes and the plainly insane.

It hardly need be said that both the Enlightened and the Romantic views of Man are still very much alive, e.g. in the works of the liberal Szasz and the existentialist Laing. It is therefore sufficiently interesting to criticize them. The easiest is to undermine the philosophy of science underlying the Enlightenment's philosophy of Man, and the social philosophy of the Romantic movement, since the chief concern of the one was scientific progress and of the other was the social order and the overcoming of Man's alienation.[19]

Modern social philosophy is almost entirely polarized into individualistic (18th century, Enlightenment) or collectivistic (19th century, Romantic).[20] Most psychology (the exception being Jung, who is not relevant in this context) and individualistic sociology, especially the school of Max Weber, belong to the first kind.[21] The trouble with them is that they have no room for deviants, or, at a pinch, they have to lump all deviants together. It is possible within individualistic sociology, to describe the paranoic, or the prejudiced. But neither will be deviant, or else they will both be deviants in the same way. The reason is very simple: the individualistic model does not have society as a separate entity with its own separate norms from which the individual can deviate. The only ways given to an individual to deviate within the individualistic philosophy is a recognizable peculiar defect, such as ignorance, error, physical defect, etc. Needless to say, error is the prime defect within individualistic psychology or sociology, and so most important deviations are errors and most important deviants are those who hold to given errors tenaciously and systematically enough to be recognized.

The second kind of modern social philosophy, the collectivist one, which begins with Romanticism, has been entrenched in social science by Emile Durkheim and the functionalists.

What characterizes collectivism is the idea that society has its own norms and imposes them on the individual, whether he likes it or not. Now in many collectivist traditions the clash is between society and the individual. Romanticism saw some value in that clash for society, since it allowed for progress. Later collectivists went further. Karl Marx saw in the individual who breaks from the norms not one on his own but rather a representative of the norm of a segment of society, to wit, a class;[22] the clash then became one between segments of society, not between individual and society.[23] The value of this change was in the making of the individual hero, not some arbitrary diviner, but one whose conduct is explicable by some norms.

This idea was generalized further by Durkheim and the functionalists: every individual's conduct is explicable by some social standard. Hence there is no real deviance. According to Durkheim, as long as society functions, it allows for no deviation, and is no real deviation, Suicide and crime are examples which Durkheim offered. The criminal, he said, serves society by reminding its members of the law, and of its importance.[24] Likewise, E. E. Evans-Pritchard[25] and Max Gluckman saw in feud a custom contributing to social cohesion. The result of all this is what Dennis H. Wrong censured as 'The Oversocialized Conception of Man in Modern Sociology', in a paper with this title, published in the *American Sociological Review* in 1961, and which received ample attention since: collectivism allows for no deviation whatsoever.[26]

Nowhere has this dual deficiency of social science been made so obvious as in the studies of Erving Goffman,[27] Wrong's classmate. Psychologically, the psychotic is one thing, sociologically quite another. Sociologically, for example, he may be in a closed institution and so resemble a Kibbutz-member or a small college professor; or he may be isolated and resemble an artist or a hermit. Psychologically, however, the artist or hermit may be a highly integrated and well-adjusted individual, permeated with social consciousness, quite unlike the psychotic.

Hence, a sane person in a mental institution, yet not on the staff, is a deviant because he is sane. A number of experiments were made, of implanting such deviants, namely sane people as patients in mental homes.[28] Needless to say the fraud was not detected. This is usually explained by the claim that the diagnostic tools of the hospitals were poor. More probably they were geared for psychology, not for sociology. The doctors were trained but not to detect deviation; they were alerted but not for that task; and so they did not perform it at all. Significantly, patients could do that with ease: they diagnosed the plants as deviants, and concluded they were sane. It is beyond our credulity to believe that the doctors could not do the same, yet this is what investigators repeatedly report with regret. Now since it is agreed that ordinary diagnosis of paranoia does contain a sociological component, and since psychologists are seldom sociologically trained, this gives us an insight into the terrible blindness that diagnosticians are prone to daily, and so we do not wish to contradict the criticism of the diagnostic tools. Yet we do think it is unfair to the psychiatrists who did their psychological jobs better than the literature credits them for, and that consequently the discussion goes in the direction of increasing disregard for the sociological aspect of the picture.

The upshot of all this is that we cannot have a purely psychological definition of paranoia which will satisfy us, as long as we agree – and we all seem to agree – that paranoia is partly psychological and partly sociological,[29] that is to say, as long as we agree that the mental make-up of the paranoic brands him as a deviant in some sense. Now this complicates matters unbearably, and for a very obvious reason. Since some paranoics are in mental homes, some not, and, in each case, some rightly, some wrongly, it is hard to say in which sense the paranoic can at all be described as a deviant. Things might look different if social philosophy had offered us a cogent theory of deviation. But, as we have said, it does not.

This is not to say that there is no sociology of deviation whatever; at the very least we do have criminology which surely is a science of deviation. Yet all that the sociology of deviation tells us is that the deviant may be marginal to his society, and the major role of criminology is to condemn our out-dated protective and penal system. With all due respect, we can ignore all this in the present study.

It might be tempting to conclude that paranoia should be defined not absolutely but relatively to any given society.[30] We shudder at the thought of this: our sociological knowledge is so scant, that to make psychopathology depend on sociopathology will delay our understanding of mental illness almost indefinitely. In our theory, which we will present below, there will be, in a compromising measure, some relativization of paranoia to the paranoic's society, but not beyond what is the most rudimentary and commonsense social knowledge, and such that is uncontroversially accepted by common members of that society.[31] Moreover, we shall later claim, an essential ingredient of the paranoic's defect is intellectual and gross: he ignores certain views and norms which are public knowledge; he is a deviant[32] in a manner he can hardly be ignorant of, yet he manages to be ignorant of it. But more on this later.

CHAPTER 4

METHODOLOGICAL BACKGROUND

It is with some measure of hesitation that we include this chapter in the present study. Though its point is crucial to us, it may be useless to some readers. It is, indeed, so crucial to us that we think this is the point which has enabled us to have arrived – as we think we have, rightly or wrongly – to the place where the greatest contemporary lights have not. We think the problem we set to solve, namely the resolution of the paradoxes of paranoia, is so intricate that it easily eludes anyone who adheres to traditional views on scientific method. This is not the only problem that traditional methodology has blocked the solution to. Indeed, many problems in the field were solved while ignoring traditional views on method. As, for example, when Freud refused to accept the verdict of the evidence for a while, contrary to the canons of scientific method (and he was right: the canons are too stringent, of course). Our problem, however, seems to require more than the mere violation of the accepted standards: it requires a new and more sophisticated methodological apparatus.

Nevertheless, we have met with severe censure from psychiatrists who hold more conventional views, or who merely play it by ear. This is not the place to try an extensive examination of methodological issues, and even our report on the goings-on in the field of methodology must of needs be very sketchy and cursory. Nor is this the place to provide psychiatrists with the new ideas and techniques recently developed in methodology. Trial and error made us choose the following presentation where, to begin with, we offer a very brief historical sketch tailored to our specific needs. We have no intentions to convince the reader that our methodological views are correct, though, of course, we happen to think they are. But we are of the opinion that if the reader takes the present chapter as a mere reflection of our method and methodology it will be easier for him to see why we chose to take the paradoxes of paranoia so very seriously, as few writers do, and attempt to offer them a clinical solution rather than the sociological solution Laing and Szasz offer. In brief, this chapter, right or wrong, presents a methodology that we try to apply in the rest of this volume, and reading it may help see the lines of development, we hope, exhibited here. Readers who find it heavy going may wish to skip it, at least at first reading. Skipping this chapter will not interfere, we hope, with the reading of later chapters.

The major problem of methodology is, is scientific method possible? Scientific method, or a method of discovery, or an algorithm of discovery, or a science-sausage-machine, is a mechanical or almost mechanical procedure, with a fairly clearly specifiable input and with science as its output. It is a strange fact that from the middle of the 17th century to the end of the 19th century, the positive answer to this question was most popular, that in our century the negative answer to the question is equally popular. Yet, this being so, it is perhaps not very surprising that we often slide back; that, nevertheless, unnoticingly many people now and then find themselves still clinging to the affirmative answer. For example, J. Robert Oppenheimer[1] clearly and emphatically denied that there exists an algorithm of science, a science-sausage-machine; yet he was convinced that sufficient Federal funds would permit the collection of sufficient hardware and brain-stuff of sufficiently high quality to ensure scientific progress (the accent is on the word 'ensure').[2]

Philosophically, of course, the input of the science-sausage-machine is neither money nor laboratory hardware, but human endeavour. What this endeavour is, was classically contested between the two leading schools of thought, the empiricists and the intellectualists. The former prescribed pure empirical findings – lots and lots of them – and sparse theorising based on these; the latter prescribed clear and distinct thinking plus logic. These are the ideas of Sir Francis Bacon, the empiricist, and René Descartes, the intellectualist. Both promised results. Both had to explain not so much how they were sure to get results, but how come the powerful science-making machine was left unemployed until they appeared on the scene and advised people to employ it. Given that scientific method exists, how could it be left to idle for so long?

The already discussed theory of errors – Bacon's doctrine of prejudice – explains just this puzzling phenomenon. It is not that the sausage-machine was left to idle, but that it was misused and abused; it is not that when wrongly used the sausage-machine stops working; on the contrary, it goes on as furiously as ever; but the output of the science-making machine when there is wrong use or wrong input is not science but pseudo-science; the products of scientific method when wrongly used are not science, but superstition and prejudice. Bacon was convinced that abuse of scientific method enables one to use empirical evidence with ease to the most unscientific purpose, namely that of entrenching error in the name of truth.

The proper input of the science-sausage-machine, said Sir Francis Bacon, must be pure facts and the purity of the facts must be jealously guarded. Once we get negligent and add a theory to the input, we are sure to get it in the output, and confirmed as if it were true. Still worse, once we color the facts with the tincture of a theory, say by accepting the emphasis it places on given facts or even on certain of their aspects, then we are sure to obtain that theory in the output of the science-making machine which has to process these facts. Similarly we can be impatient with scientific method and wish to obtain a theory prematurely, i.e. before we have enough facts to obtain it in the proper manner. Again, our wish is granted and we obtain a theory and the theory looks scientific. But it is not. Dr. Johnson's madman (see chapter 3, note 14), Freud's neurotic (chapter 1, note 10), Kraepelin's paranoic (chapter 5, note 14), Ginsberg's antisemite (chapter 3, note 13), and even Popper's dogmatist (chapter 5, note 2), each of them is but a variant of Bacon's inherently rational man gone somehow astray and become unnoticingly irrational – lost his reason due to carelessness. In this they are all modelled after Bacon's bad scientist, who, instead of waiting for the facts to lead him to a theory, dares invent one and test it empirically. Clearly, said Bacon, his test will not be of any use, since he will not give up his theory just because a small fact contradicts it. This, added Bacon, is especially true if he has disciples. He will have a fixation on it; he will rather distort *ad hoc*, either his theory or his facts, than give up his pet doctrine and the advantage it gives him over his disciples. And, unable to take the cure of accepting facts which run contrary to his theory, he will be trapped in his error; his theory, thus, will act as both spectacles and blinkers.

In parenthesis we may examine Bacon's idea that one's disciples increase the force of one's fixation on one's doctrine. It is not the idea that one's company lends sanity to one's views. He did not speak of insanity, but of the nearest to it he knew, namely of superstition. And he was quite willing to condemn off-hand as mere superstitions all views which were extant at his time. If anything, this would class him, perhaps, as paranoic (except that he was fully aware of it and so quite sane, if somewhat hysterical, perhaps[3]). Interestingly, he stated quite clearly that the desire to dominate people's minds is very strong and offers an incentive to speculate and develop schools of thought.[4]

It is quite clear that for Bacon it is essential that scientific theory should be demonstrably true, since error acts as spectacles and blinkers and so inevitably becomes pseudo-science and prejudice. And so, in order to know the difference, every theory must be demonstrable before it is seriously offered or even merely entertained.[5] For, anything short of a complete proof may be a delusion. And obviously, the nearer to proof a delusion can come, the harder it is to detect it, and so the more dangerous it may be.[6]

Bacon took very seriously the integrative function of science as well as the integrative function which an error may assume when it becomes illusion and pseudo-science. He argued that in science the chief purpose is integration. The proper process of integration, he said, is slow but sure. Also it is a process which culminates in a completely integrated world-view, a scientific metaphysics. An error, on the contrary, operates at once as an integrative principle: it becomes increasingly central, the more its holder sticks to it in the face of facts: it develops increasingly the semblance of scientific certainty and so finality. Indeed, he added, the quick and large returns of the method of speculation is what makes it so alluring. It is hardly necessary in this essay to draw attention to the perceptiveness Bacon showed here.

Descartes agreed with Bacon's theory of error, but doubted that facts alone can ever be reliable. To make them so, one must begin with sure integrative principles, with scientific metaphysics. We need not dwell on his views here, since, historically, Newton was viewed for a century or two as Baconian and as a prime scientist who could make no mistake whatsoever; and so, traditionally, Bacon's view dominated.

There is no doubt today among leading methodologists and leading historians of science that scientists always ventured hypotheses. During the Baconian era they were regularly frightened by this fact. From time to time, the practice of making very tentative hypotheses was justified as a cautious preliminary to the achievement of scientific theory which, when exercised with much discretion, may be quite benign. The most successful attempt in this direction was the work of Dr. William Whewell, around the middle of the last century. Understandably, his work had enormous influence on scientists, yet he was soon forgotten, to be discovered only after World War II. He was forgotten, although he believed that scientists establish absolute certitude subsequent to their suggestion of tentative hypotheses, because he gave up the idea of science-sausage-machine, or of a science algorism. Whewell described science as exclusive of error by marking sharply and clearly from each other two kinds of error. Quite apart from the body of rigorously proven hypotheses, i.e. science, he said, there are tentative hypotheses and there are shakily proven hypotheses, i.e. pseudo-science. Whereas Bacon believed that all tentative hypotheses becomes pseudo-science, Whewell believed that tentative hypotheses may be severely tested and scientifically verified or refuted, or alternatively crudely tested and pseudo-scientifically verified. This idea of Whewell's equates the pseudo-scientist with anyone who is sure of ideas not yet publicly proven (and by remote implication with the paranoic); it is still very popular: it is one thing, they say, to venture a guess but keep an open mind on it, and quite another to view it as science proper even though it is still unproven. Moreover, viewing one's pet hypothesis as scientific naturally leads to a (paranoic) persecutionist view of the world of learning as explanation of its unjustly withholding scientific recognition. And so, strangely, Whewell, though rather forgotten, is still very influential in scientific circles, especially in psychiatric ones.

There is here a great difficulty which was never noticed, perhaps because it was deemed more hypothetical than real: when one proves that a theory is true and communicates it to the world of science, at that very moment one stands apart from the world of science and may be viewed as a pseudo-scientist and paranoic. Is he in such a predicament as long as he stands alone? Of course, this is a mere teaser, because we normally take our time in order to see how the world of science responds before we make up our minds. And while the world of science takes time to deliberate, we simply suspend judgement. At times, naturally, the world of science is receptive to the new idea and thus proves its proponent eminently sane. At times it rejects the new idea in a manner which makes its proponent relent, thus likewise proving him eminently sane. At times, of course, the scientific world may be regrettably prejudiced and until it comes round to accept the proof of the new theory as offered by its proponent, the proponent, though a proper scientist, may get desperate, lose his mind in anguish, or commit suicide (Cantor, Semmelweis). But, nevertheless, it seems, the problem remains rather academic.[7]

We cannot judge how academic the question sounds today, because it is based on presuppositions which have been meanwhile relinquished: since the Einsteinian revolution in science, certitude has been relinquished by most philosophers of science. Yet, clearly, even if it legitimately ceases to concern those whose sole concern is with scientific method, nevertheless for the study of paranoia it may be of great interest. If, as we shall argue later, paranoics stick to their private theories with a semblance of utter certainty very much the way Bacon or Whewell describes, and if our culture will soon be so imbued with the recognition that even in science the quest for certainty is chimerical, then perhaps even in the near future possibly this change will entirely alter the picture of paranoia in modern societies. Perhaps not in the least: this must await further clarification and observation. Meanwhile let us continue with our discussion of scientific method and pursue the difference between science and insanity regardless of how up-to-date views of this difference may in the future affect the symptoms of insanity.[8]

In our days, when certainty is gone, and with the science-sausage-machine *a fortiori* gone as well, it is hard to imagine the violent opposition which Whewell's philosophy encountered when it was proposed, a century ago or so. For Whewell science begins with hypothesis, and so it contains a risk element: we may be unable to guess a new hypothesis, or, at least, we may never hit on a good hypothesis, and so success is not guaranteed. In short Whewell was branded an intuitionist. This was enough reason for philosophers like John Stuart Mill and his followers to dismiss Whewell.[9] Nevertheless his ideas had a great appeal to men of science since he reinforced the accepted view of science as certainty — as illustrated by the exalted status of Newtonian mechanics — and since he encouraged scientists to excite their imagination — as illustrated by the flair of electromagnetics, of evolutionism, and of geology. After the Einsteinian revolution, after Einstein offered a better alternative to Newtonian mechanics, even finality had to go by the board, and only imagination remained as a necessary ingredient in the process of developing theoretical science: Newtonian certitude was the last instance of scientific certitude ever offered by any philosopher.

The question then is, what makes science special? To use Russell's imagery, what makes the pronouncements of Einstein any different from the ravings of a madman?[10]

There are three answers known which gave rise to four schools or traditions of thought, two classical and two modern: the inductivist (which Russell endorsed), the instrumentalist (Poincaré, Duhem, the pragmatists), the critical (mainly Popper) and the post-critical (mainly Polanyi). According to the inductivists, a scientific theory is made certain or at least probable on the basis of known facts, whereas the raving of the madman is improbable or even palpably false.[11] According to the instrumentalists, science is blessed with mathematical elegance and simplicity, whereas the madman constantly gets entangled with even the most obvious facts and must regularly add *ad hoc* and *addenda et corrigenda* to everything he says.

As for the critical and the post-critical schools, they are both still in early stages of development and so it is hard to give a concise summary of them. But let us present the views of Popper and of Polanyi as briefly as we can.

Sir Karl Popper, among philosophers the founder of this school (Einstein may count as its true founder), though he quotes Russell's question about the difference between science and madness, does not answer it fully. He does say what he thinks science is, and even what metaphysics is, but not what is the status of the ravings of a madman. Scientific theories, says Popper, are highly testable, namely highly refutable, namely they can be empirically shown to be mistaken, if they are mistaken. Pseudo-scientific theories, says Popper, are untestable, yet presented as if they successfully passed severe tests. Alternatively, perhaps (to use an example of Whewell which Popper would accept) they are testable, tested, refuted, corrected *ad hoc*, and then presented as if they have passed the test. This accounts for the epithet 'pseudo'. In particular, this will be justified by Popper's view that *ad hoc* corrections reduce testability – a fact which we shall use later on. As to metaphysics, it is simply irrefutable. (At times, Popper equates metaphysics with pseudo-science. The epithet 'pseudo' is evidently unjust here because metaphysics had a pretence for certainty only at times, and never empirical grounds.)[12]

Whereas Popper's view has a clear affinity to Whewell's, Polanyi's view is a mixture of traditionalism and instrumentalism. In his view science is first and foremost a community of people who share an activity — a professional community. One who dissents from the whole scientific community for long without managing to sway public opinion is too alienated to count. Yet when the integrated opinion of the scientific community becomes cumbersome and outdated, change may be effected. Once a new integrated scientific opinion has evolved, men of science can proceed with their researches in accordance with it. This formulation of ours is somewhat indebted to the reformulation of Polanyi's views by Thomas S. Kuhn; yet, whatever Kuhn has added to Polanyi's system we overlook here. In any case, Kuhn's greatest debt to Polanyi is in the idea that science cannot be demarcated abstractly, as a system of ideas of this or that kind, and must be viewed sociologically as a community sharing an activity and a view of the world, though subject to change from time to time.[13]

It would seem that the two older schools, the inductivists and the instrumentalists, have the upper hand here, as they find it easier to demarcate science. From the viewpoint of the present essay, which is more concerned with the demarcation of madness than of science, this is no advantage at all. Indeed, both the inductivist and the instrumentalist equate all error with pseudo-science, as we saw, be the errors ideas of charlatans or madmen: for the inductivists all unscientific views are improbable or false, for the instrumentalists they are all much too complicated. And so, these two views are too simplicist. They are too simplicist even if we view madness as the extreme: to say that madness is extreme error or extreme complexity is still too simplicist. But, and much worse, both inductivism and instrumentalism are too simplicist even from the viewpoint which concerns itself with science only, as they cannot account for the accepted view, which belongs to Einstein, that a new view should explain the success of its predecessor (or predecessors) by considering all previous successful theories as special cases and as first approximations. Popper's theory of science as testable explains this at once, on the assumption that degrees of testability increase with the degree of explanatory power. Those who (rightly) question this assumption may reform Popper's definition of science as the body of testable hypotheses to a definition of science as the body of testable explanatory hypotheses. One way or another, no doubt, considering a successful predecessor to a given hypothesis a special case would make the hypothesis an explanation of the predecessor's success, and making the predecessor a first approximation suggests ways of crucial experiments, which of course, are excellent tests for the new hypothesis. And so, it seems clear that the critical view has a great advantage both as a theory of science and as one which need not lump all non-science together. (Popper does tend to lump all non-science together, but his philosophy allows one easily to show more flexibility.) A similar idea should work for the post-critical school, who view the task of scientific research as increasingly sophisticated. They have not yet worked out their theory of sophistication in research, and it is not too clear how the idea of increased sophistication accounts for the (Einsteinian) idea of scientific theories as successive approximations. The problem is intriguing, and it will not do to say that series of successive approximations are series of increased sophistication: sophistication is a quality of a closed-view, of a coherent view which the world of science holds at the time it develops solutions to specific problems in the form of specific theories that approximate their successors.

And so, obviously, both the critical and the post-critical views of science are problematic. This may amply justify the majority of philosophers in their reluctance to give up the older inductivist or instrumentalist view of science and join the *avant-garde;* it may even explain the return of some to the old views. For our part we do not see the possibility or attraction of going back to older views.

This is not the place to explain the failure of inductivism and of instrumentalism to take into account the Einsteinian view of science as a series of approximations of the truth. Of course, if we define probability so that, whatever it may be, the probability of a better approximation is higher than that of a worse approximation, then inductivism is thereby fully vindicated. But within the inductivist tradition the word 'probability'[14] is used in the sense of the mathematical theory of games of chance. And, according to that theory, a hypothesis which is more general is necessarily less probable than the less general one, e.g. any of its special cases. For example, 'all my cards are spades' is less probable than 'all the royal cards I have are spades', and the latter is, of course, a special case of the former (in the technical sense of the term 'special case'). And if inductivism requires that the new theory is both a better approximation and more probable in the light of evidence, then, at best, the second requirement is quite redundant, not to say in conflict with the first.[15] As for the instrumentalist, he has no need to correlate the theories of Newton and Einstein: he views each as elegant and useful in its proper place. Indeed, the greatest attraction of instrumentalism is that it declares both Newton's theory and Einstein's theory quite valid – each in its domain of application – but not comparable to each other. (Here the instrumentalist and the post-critical agree.)

Methodology is these days in a state of flux, the two dominant traditional doctrines being unsatisfactory and the two new ones (incomplete, if not also unsatisfactory) slowly taking over. We shall not have, then, to go further into details of the critical and post-critical methodology, very tentative as these surely are. (Tentative not only in the sense that each of the two may, of course, be superseded, but even in the sense that they are still in the process of being thought out in their detail and so not yet fully formulated.)

What we wish to employ in the present chapter which belongs to methodology proper are some guidelines for the rest of this study. The first concerns Einstein's rule that a new theory should explain the explanatory success of earlier theories by presenting them as first approximations. Einstein's rule hardly applies to the meager body of theories on paranoia, where theorizing has had so little explanatory success thus far. Nevertheless, we shall attempt to present earlier theories of paranoia as approximations to ours. Second, following all the methodologists in the field, we shall view *ad hoc* corrections to given theories as undesirable though at times not entirely avoidable. Third, unlike most or all of them, we shall not endorse any hard-and-fast rule of demarcation of either science or pseudo-science or madness. This, naturally, only widens the possibility that the paradoxes of paranoia are here to stay. In our theory, however, we shall offer a complete resolution of the paradoxes without depending on any precise demarcation. Indeed, our resolution will make place for the fact – and we report it as a fact – that a regular though small part of the modern population are both active scientists and ambulatory paranoics.[16]

Finally, there is one most exciting aspect of methodology which, incidentally, greatly enhances the paradoxes of paranoia as it relates to integrative principles or closed views. The classical view of science required that anything to do with science should be certainly true. The instrumentalists eschewed truth but required a low toleration for *ad hoc* adjustment. It is not clear how low is low, and it is not clear how integrated science must be at any stage. This criticism holds for the post-critical philosophy. As to the critical philosophy, it either disregards integrative principles or has trouble relating them to high degrees of testability or is being worked out in detail.

The importance of this point is only now emerging. It was Émile Meyerson who began the trend early in the century, describing metaphysics as the integrative principle, or sets of principles, for the physical sciences. Alexandre Koyré and I. B. Cohen, both historians of science, followed his footsteps. In the meantime E. A. Burtt developed similar ideas that are regaining popularity.

We shall discuss all this in our next chapter. Let us only repeat that only when certainty is required of an integrative principle does Kraepelin's condemnation of the paranoic's erroneous principle show any degree of plausibility.

CHAPTER 5

METAPHYSICAL BACKGROUND

We have left a point in the middle, a few paragraphs back. Almost everybody, we said, confuses pseudo-science, metaphysics, and madness.[1] Leaving madness aside for a moment, as a topic which we shall soon study in detail, we wish to present here the theory of the role of metaphysics in science as an integrative principle. We shall do so briefly, and from a very specific point of view. We ask the reader to keep in mind that in the present study our concern is in the paranoic's use of his integrative principle rather than in the scientist's similar use. Of course, we do not see this as licence to distort the scientist's case or even to exaggerate: only by presenting matters rigorously can we hope to impress the reader with the force of the paradoxes of paranoia – or rather of the fourth paradox (p. 5 above), concerning the role of the paranoic's *idée fixe* as an intregrative principle.[2] It would be natural, then, if in an exposition of the role of metaphysics as an integrative principle in science we should dwell on what will come in handy in the study of the parallel case in paranoia.

Consider Darwinism. When we wonder about a rabbit's long ears, Darwinism, as an integrative principle, will suggest to us to propose the following historical hypothesis. Once upon a time two strains of rabbits competed for one ecological niche too small for both; one strain had ears longer than the other, and that one strain had an advantage; and so it won. The other strain, then, had to disappear. Now, this hypothesis can be further explained by an additional bio-physical hypothesis about the acoustic advantage of long ears. Both hypotheses are refutable. In particular, long ears may be no advanage at all, and even an ecological disadvantage, yet it may be a characteristic which is coupled genetically (i.e. in the specific make-up of the genetic material of the rabbit) with some great advantage as yet unnoticed.

Consequently, we have here a specific possible explanation or even a set of specific possible explanations, each of which in principle is testable, all suggested by, or conforming to, one integrative principle. This connection, the conformity of the hypothesis to Darwinism, is a point easier to comprehend quite intuitively than to formulate with a sufficient degree of precision. Clearly, the Darwinian integrative principle itself is viewed as important because it is integrative, and so we do not mind too much the obvious fact that it is itself not empirically testable. Nor will it become testable before we have a complete set of alternative hypotheses conforming to it, all explanatory to a given set of facts, and each of which is independently testable. For, only then can we ever hope to exhaust the possibilities it offers and eliminate them all empirically and thus declare it, the principle itself, empirically superseded. Usually, metaphysical principles get superseded in a much simpler way by being replaced by better ones.

The easiest example for the rare case of a testable integrative principle may be Daltonian chemistry. We shall not illustrate it since it is by now too obvious to anyone sufficiently familiar with high-school chemistry. Suffice it to say, however, that for long it stood, as formulated by Dalton himself, aloof from empirical tests and yet was highly valued, because of the fact that it was presented as a principle generating empirical hypotheses; it repeatedly stimulated scientists to propose an ever increasing variety of chemical hypotheses about atomic ratios and structures in given molecules, which hypotheses were tested empirically, and often with significant results. Yet, finally, with nuclear disintegration, with chemical isotopes, with endless molecular chains, this integrative principle, Daltonian chemistry, was refuted too, and so it was superseded and complemented by nuclear chemistry etc. Take the theory of the radical (such as CO_3 or SO_4) as a part of a molecule with strong interaction or affinity between its atoms and so stable enough to behave like an atom in the face of not too strong chemical upheavals. Even this theory acted as an integrative principle, within classical chemistry – within organic chemistry, in particular. It was refuted by Kekule's discovery of the benzene ring, which, although made of carbon and hydrogen, acts as a radical in view of its structure rather than in view of the affinity between the two. Yet, the radical theory was immediately revised and continued to act as an integrative principle. Indeed, up till today, we cannot present a very heavy complex molecule, such as an amino acid, without presenting it as a complex of atoms and radicals, and where the radicals are units of limited stability, whatever the cause of that stability.

In the classical days of science, when it was customary to view science as certainty, the status of any integrative principle was quite a serious problem: when in doubt it was deemed extremely dangerous, and when certain a great boon. We have discussed this before, when presenting Bacon and Descartes: Bacon warned against all integrative principles, knowing full well their allure to researchers and fearing that their ill effects will spread to distort our views of all sorts of detailed facts. Descartes thought he had invented scientific metaphysics, an integrative principle immune to the danger of error. The modern theory of metaphysics as integrative principles in science in general is due to Émile Meyerson.[3] What we have to do here is to integrate Meyerson's idea with some modern view of science – and we only have two such views, the critical and the post-critical, or, more specifically, that of Popper and that of Polanyi. This will offer us the most advanced theory of science; our concern with it here, however, is of a mere preliminary to a theory of paranoia. We shall take up the critical view first.

Popper's theory says nothing about the generation of hypotheses – except that they cannot be generated by an algorithm, namely automatically. (This, we remember, is by now universally accepted as a matter of course.) When hypotheses are generated, he says, our job is to test them. Also, given a few competing testable hypotheses, he suggests we choose and examine first the most highly testable one. All this seems to us open to certain modifications. First, there is Meyerson's objection: though hypotheses cannot be generated automatically, an integrative principle often facilitates the generation of a hypothesis. Second, if we ever accept an integrative principle, it may lead us to disregard the hypothesis not conforming to it, however testable it may be, and prefer one which does. Evidently, then, the integrative principle is both a spur and a constraint, and so obviously a significant factor in the decision about the direction which research may take.

An example of a testable theory rejected by science on metaphysical grounds is the continuum theory, which is regularly viewed as mere applied mathematics. It is atomism which makes us disregard hydrodynamics as a theory of fluids, and elasticity as a theory of solids, etc., since for hydrodynamics fluids are continuous and for elasticity solids are. That often enough engineers use hydrodynamics and elasticity as instruments is no argument against atomism; rather it is another argument against instrumentalism; it is likewise a strong argument against Popper's demand for high testability. It shows that Popper is in error when he views testability as a mark of realism and a blow to instrumentalism: we may test a theory either in our search for the truth or, and at times quite separately, in our engineering ventures.[4]

This is not to say that critical philosophers must demand that integrative principles should be endorsed uncritically. On the contrary, since certainty is already gone, critical philosophy may notice that within one science competing integrative systems are allowed, each propounded by researchers who use it in order to propose testable explanatory hypotheses conforming to it. And, of course, this might very well make researchers train themselves to switch integrative principles at will, so as to make them able to compare a few before making a choice; to make them able, that is, to examine different integrative principles critically before making a choice.[5]

Admittedly, integrative principles are too often endorsed uncritically, and not always as much subject to examination and deliberation as they ought to be if scientific progress is to be served as efficiently as possible. Worse still, there is a simple technique by which to make an integrative principle accepted all too easily and it is as follows. A hypothesis conforming to a principle which explains a given fact may be presented either as a part of the principle, or as the fact, thus supporting the illusion that the given fact is explained by the integrative principle itself, perhaps even that the fact is the outcome of a successful test of that integrative principle which is thus (allegedly) empirically supported. Obviously, in such a case the integrative principle degenerates into pseudo-science exactly in Popper's sense. It is no accident that Popper's chief examples of such pseudo-scientific theories are the integrative principles of Marx, Freud, and Adler. Popper argues that Freudian or Adlerian theory are pseudo-scientific by the following example. If a man throws a child into the river, individual psychology will predictably interpret the fact as an act of self-assertion; if a man jumps into the river to save a drowning child, individual psychology will again predictably interpret the fact as an act of self-assertion. Hence, says Popper, facts can only support but never refute Adlerian psychology. Further, whereas Freud will view self-assertion as motivated by the sex impulse, Adler will view the pursuit of sex as a mere act of self-assertion. Hence, both theories accord with all conceivable facts; hence, both are untestable. Still further, if you accept Freudian theory, all the better; if not, your very resistance, since it is expected in the theory, is its confirmation. Likewise, if the capitalist press opposes Marxism, this is as expected; if it on occasion concedes a point to Marxism, that is the result of inability to resist the force of truth; and if it does neither, it is detracting public attention from the class-struggle – again as expected. (We think clinicians treating psychotics are all too familiar with this.)[2] Contrary to all this we suggest viewing both Freudian or Adlerian psychology, as well as Marx's sociology, not as testable hypotheses or as ones claiming any empirical support, but as competing integrative principles: as highly suggestive of hypotheses about the individuals and societies in question, which are themselves at times empirically testable, at times not.

We have touched here upon a very sensitive point. Any definition of science, given by one who values science and opposes dogmatism, is likely to dissociate the two and thus conflict with the admitted fact that some, if not very many, men of science are dogmatists. Indeed, Polanyi and Kuhn require this to be the case, as we have noted; we shall return to this later. But Popper has to admit this as a fact concerning science conflicting with his definition of science. Popper can easily escape this conflict and say his definition is a kind of spotlight that brings to our view what is good in science. Agreed. We can do the same with integrative principles no matter how dogmatically they have been held by so very many scientists. And certainly Freudianism, Adlerianism, or Marxism, can be viewed in the same way and their contribution to the formation of testable scientific hypotheses can be examined. The same can be said of many other integrative principles, and in the following manner.

The most important, most general, integrative principles of psychopathology are either the view behind the organicist[6] approach – outside professional circles it is identified not very correctly with mechanism – and the view behind the psychogenic approach – outside professional circles it is identified also not very correctly with mentalism or dualism. (Dualism assumes the existence of independent entities, the mental and the physical.) Now, often the labels for the approaches are prefixed with the word 'methodological', e.g. 'methodological dualism', in order to dissociate them from the views behind them, so as to stress that it is not a metaphysical commitment to a given metaphysical doctrine at stake, but a methodological proposal, to try to generate hypotheses conforming to that doctrine. There is no doubt that the integrative principles are operative and have greatly influenced, in different ways, those researchers who tried to abide by them. In this study, intended for the profession, we find no need to illustrate the point, especially not in the days when research centers and even clinics exist which operate on a strictly organicist or a strictly mentalist approach as the case may be. Suffice it to say here that all the psychopathology in the present study tends towards the psychogenic approach, though it is not dualistic, as it is not quite in accord with the assumption of two different entities. (To avoid confusion, we ask the reader's indulgence when we restate the obvious, namely, that the psychogenic approach[7] does not exclude organicist hypotheses[8] on occasion, though it more often centers on mentalistic ones, whereas the organicist approach does strictly exclude psychogenic hypotheses, and recommends that we convert any successful psychogenic hypothesis into an organicist[9] one, preferably as a special case and a first approximation.[10])

Now, clearly, looking in a benign scientific manner into the integrative principles that have been operating in science over the generations, we may have missed something.[11] If we allow ourselves to ignore the unpalatable we may raise the paradoxes of paranoia all too easily and quite uninterestingly. For, using the same idealization on both the scientist's integrative principle and the paranoic's integrative principle we bring them together quite artificially!

One might refer here to the interest, the intrinsic intellectual value, of the scientist's integrative principle, absent in the case of the paranoic's integrative principle. We are reluctant to do so since we cannot tell *a priori* that every integrative principle tried out by a scientist is worthwhile, nor would we know what to do with a paranoic who endorses an interesting integrative principle.

Similarly, we would not wish to resolve the paradoxes of paranoia by contrasting scientific flexibility with paranoic inflexibility. In the first place, not all scientists are so very flexible. Moreover, while dissociating the paranoic from the flexible scientist, we may all too easily identify him with the philosophically committed existentialist. To show this we shall have to introduce the existentialist's inflexibility, the existentialist doctrine of commitment, so-called. This will bring us easily to Polanyi's post-critical philosophy which describes scientists as committed, and so not much like the flexible cautious scientists of the classical views or the popular views – yet of course very far from paranoics.

So far we have confined our discussion of metaphysics strictly to the field of methodology, that is to say, to the study of scientific research.[12] In the popular mind, however, metaphysics is connected with life in general, or with the meaning of life in general, not necessarily the life of science or its specific meaning. Now, for those who approach life as if it were to a large extent one big experiment, and who also consider science as primarily experimental, for them this contrast between the life of science and life in general hardly matters. But too many people will object to this dual approach. They will object either on the ground that science is more than a mere experiment or a mere adventure; or on the ground that life is more than a mere experiment or a mere adventure; or even on both grounds. Most existentialist philosophers take life very seriously as a matter of course. They also take science lightly – perhaps as a mere adventure, perhaps as even less than that, namely as mere computation techniques for engineers (instrumentalism). For our part, we fail to see the contrast between seeing life or science as a series of adventures and seeing it as something very serious. Rather, we are convinced that at least one existentialist, R. D. Laing, will agree with us that life is a very serious experiment. Yet, most existentialists demand from people that they prove their seriousness by committing themselves for a lifetime to only one metaphysical principle – it matters to them less which metaphysical principle – quite in opposition to any experimental approach to metaphysics which may recommend that we try one metaphysical system and then another. Now, for the sake of historical accuracy we have to notice that, unfortunately, commitment is usually contrasted not with experimentation but with skepticism (even though etymologically skepticism is an attitude of search). And so, perhaps, before discussing what commitment may be, we should observe that one may be committed to – or be skeptical (in the sense of being in doubt) about – science, pseudo-science, and metaphysics. The metaphysics in question may be of diverse kinds; in particular, it may be, we remember, an active integrative principle of research, and it may be an integrative principle degenerated into pseudo-science, and so discourage further research. Or, perhaps, a metaphysics may at times be rendered empirically testable and thus become scientific. The question, in either case, is, in which way is commitment contrasted with doubt? What is the meaning of commitment?[13]

There is much hostility to skepticism which is of no interest to us – from the quarters of believers in science as certainty, from dogmatic schools of thought, and even as mere expressions of some psychopathology. Let us ignore all that. There is, further, hostility to skepticism, at least to certain classical versions of it, which comes from pens of even such philosophers as Bertrand Russell. This celebrated author of the exciting 'Free Man's Worship' and of *Skeptical Essays,* dismissed in his *History of Western Philosophy* the whole ancient schools of skepticism as facile. It is all too easy, he felt, to leave every question open under the – true enough – pretext of ignorance. In life we must make decisions on insufficient grounds.

It may be alleged, then, and not only by existentialists, that the skeptic is insincere, that his own views make no difference to him, that he is not in the least committed to act on his own views. If so, then its opposite, commitment, is the willingness to back one's convictions by action, by paying the price of one's conviction, by showing the courage of one's convictions. But this reading is an error because classically the scientific attitude, as understood by many modern authors, including Bertrand Russell, encompasses both scientific skepticism and commitment to (allegedly) properly endorsed views (i.e. to views endorsed or recommended by science). This view, that we must be skeptical until science recommends a view and then we must endorse it, is what we normally call scientific caution. And so, whether we accept or reject the classical rationalist view, we cannot endorse the reading of commitment as meaning merely the courage of one's convictions: the existentialists have certainly recommended commitment in opposition to all skepticism whatsoever, including explicitly also the classical rationalist view which recommends scientific caution.

It is here that the post-critical philosophy, especially of science, can come to the rescue with a new view of commitment and of scientific caution. Indeed, Polanyi's triumph is his theory of commitment to science.

The new theory of commitment is a theory of commitment to a way of life, and it thus applies to religion, politics, art and science. It says, commitment is not merely the acceptance of un undemonstrable principle as any hypothesis may be accepted for the purpose of a debate or some other exercise. Rather, the commitment is to a way of life, to a life-style. The person committing himself to a life-style becomes an apprentice, and he is committed only when he graduates. It is serious business, and there is not much room for deliberation, because one cannot know what it is going to look like when one is committed until one experiences the fruits of one's commitment.

The knowledge that is confined to the committed who has undergone his apprenticeship and graduated, Polanyi calls personal knowledge.

Personal knowledge is not free of error, and so it is alterable. But is only alterable step by step, and within a community of knowers. In each community the dynamics of change is specific to that community, and the dynamics of scientific change is subject to scientific caution.

So much for Polanyi's post-critical theory. It seems to us that Polanyi's argument is too powerful and too devastating; just as we cannot articulate our experiences as scientists, so – more so, if Laing is any nearer the truth – we cannot articulate our experiences as madmen. And the critical rationalist must concede that perhaps the post-critical view is true, yet he need not approve of it; he may claim that perhaps science progresses more efficiently when opinions are fluid, when attempts to articulate go as far as possible, and so on, that the impossibility of utter success should not deter all attempts to be rational, to articulate, to criticize.

There is a stand-off here. In science there are periods of turbulence and periods of relative calm; who is to say which is the more characteristic and/or beneficial? The post-critical philosophers acknowledge the significance of an upheaval; they insist that it must be confined, that the stability of the scientific community, in particular the stability of its beliefs, must be preserved. Critical philosophers think the individual scientist counts more than the whole community, especially when he manages to act, to make bold conjectures or sharp criticisms and generally stir things up.

Now here we have two or three integrative principles regarding science. Indeed they are the integrative principles of all social and political philosophy since antiquity, and here is the place to state them. In particular, we think, a quick glance at these, though in many ways too superficial and unsatisfactory, may explain the neglect of the paradoxes of paranoia, or even their very origins.

The root of all western social and political philosophy is the discovery of the diversity of human laws and custom and the subsequent dismissal of them as arbitrary and non-binding: I cannot accept one view rather than another only because I happen to be born to these parents rather than those. If there is one true human law and custom displaced by, or hidden under, the variety of traditions, that would be binding. The question, is there such a law, caused the main division between philosophers – to naturalists and conventionalists. The naturalists were so very strongly individualistic as to overrule any particular custom, law, tradition, because of its particularity that makes adherence to it so arbitrary. The conventionalists were cynics who saw nothing better than the merely arbitrary.

Soon some naturalists developed a theory of society as an organism, as a natural unity. Moreover, some philosophers declared it natural to make conventions. Both these answers left one difficulty intact: we still have the question, is there no truth to the peculiar law and custom of a given society. To this the only answer remained, for long, that there can be no truth, and the declaration that my religion is true is but a part of my traditional custom. The alternative was the quaint idea that truth is relative.

The important corollary to all this was a matter of authority, especially intellectual authority: does the individual decide what views are true or should the state or society or clan keep him in tow? The eighteenth century philosophers, the Enlightenment, the Age of Reason, accepted the idea that truth can be discovered by the individual and his holding it can be justified: thus, there is no need for arbitrariness or for the diversity of opinions it leads to: the one dissenting from the manifest truth is either a slave to some peculiar arbitrary tradition or a slave to his own whim. The opposite view saw authority in society and declared all deviants mad. As we said, this is the nineteenth century Romantic view. Even geniuses, we saw, are for the Romantics deviants and so mad – unless and until their society adjusts to them and thus rectifies their deviation, often posthumously.

It is thus not surprising that paranoia was traditionally the paradigm of madmen despite its rarity: it is a paradigm of an odd and arbitrary and thus irrational peculiarity:[14] it agrees perfectly with all known philosophies, it fits all known integrative principles.[15]

So strong, indeed, are these principles, that they are incorporated even in more modern philosophies, indeed in all of them save the critical philosophy. Consider the existentialists like Sartre, who recommend an irrational commitment to any principle, or those, like Polanyi, Thomas S. Kuhn, and Imre Lakatos, who recommend commitment to science in general and perhaps to today's science in particular, or, in addition, like Pierre Duhem, to both science and a religious metaphysics. Their views are hit by the paradoxes of paranoia all the harder. For, when we shall examine the structure of the paranoic's thought we shall see how much it agrees with what they commend as integrative principles to replace those of the Enlightenment and of the Romantics.

Yet, these principles are not good enough, and we have all over the place new theories and ideas that indicate the rise of a new viewpoint, a new critical philosophy of society and science, which gives room for both individual and society: it allows no final authority – no certainty – and therefore permits individuals to criticize society and society to criticize individuals. Now a society still stands for certain integrative principles and so it may very well be not the one that stands for the best ones; but it cannot be dismissed as merely arbitrary and be done away with.[16]

Hence, what was comfortable in the eighteenth and even in the nineteenth century is now no longer comfortable. The need to criticize older views may be felt ever more strongly.

(This, incidentally, is how a psychiatrist with a strong philosophical bent and a philosopher with a psychiatric curiosity have come to write the present volume.)

CHAPTER 6

THE PARADOXES OF PARANOIA REVISITED

We begin by considering the paradoxical axiom, which is the most widely accepted, by the profession and others, that the paranoic is in need of help though he suffers from an intellectual quirk. We can present the paradox of paranoia at once as the very well-known fact that we may observe two people with similar or seemingly similar intellectual structures (or thought-patterns, or practical logic, or what have you), one fairly independent and the other in dire need of help. Or we can present the paradox of paranoia at once as the very well-known fact that a paranoic needs help even though he may be, and often is, more intelligent, more perceptive, and more self-aware, than an average normal member of the community who can manage much more independently.

But we can even present the paradoxes of paranoia while explaining the paranoic's need for help. It is, no doubt, highly problematic. Some paranoics, it is well-known, are subclinical or ambulant: they may hobble along fairly reasonably, and manage to contain their enormous fears and sufferings, or perhaps they carry on whilst repressing these in a surprisingly stable fashion keeping them underneath a thin surface, or perhaps they keep their volcanoes under control by engineering now and then small and manageable eruptions – clashes with kin, neighbour, colleague, or on occasion even the police. Be it as it may, the very existence of sub-clinical or ambulant paranoics makes even the most obvious axiom – that paranoics need help – quite questionable. It is not surprising, then, that some will not class sub-clinical or ambulant paranoics as paranoics proper, saying they are not really ill. This, alas, will not solve our difficulty.[1] Suppose we declare the sub-clinical paranoic as (well and) non-paranoic. Now, undoubtedly, what may bring a sub-clinical or ambulant paranoic to the clinic and thus qualify him as a paranoic proper may often be sheer accidental added emotional pressure, or an eruption due to a sneak moment of truth, or a clash with a neighbour or policeman (or a mere car crash) which happens to be a jot more severe than carefully planned. Now whether this accident makes the sub-clinical paranoic into a paranoic proper, into a clinical case, may well depend on our view of him as paranoic proper or not in the first place.[2] And so the problem is not solved by a mere convention of what counts as paranoia proper to begin with.

If we assume that the sub-clinical paranoic is not a paranoic proper, then the paradox of paranoia is all the more enhanced since this means that, contrary to popular prejudice, a madman may be identical in his thinking pattern with a sane man: the same person's thinking pattern is hardly altered during the breakdown, be he a sub-clinical or an ambulant paranoic to begin with or not. And so, if we assume either that the sub-clinical or ambulant paranoic is a paranoic proper or not, we end up in trouble. Likewise, if we assume that such a paranoic needs help or not, we end up in the same place. If he does, then our paradoxical axiom is reestablished; if he does not, then our paradox is enhanced, since we may conclude that what makes a madman mad is not his thought pattern but his manifest need for help.[3]

And so, whichever way the cat jumps, we may still assume that the paranoic is in need of help and for the sake of our discussion ignore the one who does not, on the understanding that taking this course does not increase the paradox but may indeed decrease it. The other benefit of this course, an added benefit as it were, is that we face paranoia as a human problem: our patient is suffering and this is our prime concern.

Here we clash head on with R. D. Laing, the prime exponent to date of the paradoxes of paranoia (though, to be exact, their exposition *qua* paradoxes is more due to Ey and, moreover, Laing talks of schizophrenia, not mentioning paranoia explicitly). Laing goes so far as to claim in places that the psychotic is more of an integrated human being than the conformist who brands him as mad. Now, Laing is fully aware of the patient's suffering: this is his very starting point. In particular he observes the complete – emotional as well as intellectual – dissociation of the schizophrenic from his body, to be the prime symptom and problem of the patient. Indeed, Laing's sharpest statement of the paradox is where he states (*Divided Self*, Chapter 1) that Descartes as a philosopher expresses dualism, the view common to many schizophrenics, of the distinctness of body and self. Nevertheless,

Laing thinks this Cartesianism is largely the result of the aggressive conduct of so-called normal people towards psychotics. (Of course, those who have lost their freedom resent the freedom of those who dare retain it, and so they cruelly push hard the free spirit who becomes a poor patient tending towards Cartesianism[4].) Partly Laing ascribes the suffering to the very freedom – creative and emotional – of the patient, not as a suffering of a patient but as the suffering of a free agent *qua* free agent. He insists on the patient's overall moral and intellectual superiority, thus, finally, declaring the pain a small price to pay. His famous slogan means: I wish I could convince you it is a small price, with such vividness that you will breakdown voluntarily. "If I could turn you on, if I could drive you out of your wretched mind, if I could tell you I would let you know."[5] It is hard to take this weakness of Laing as more than mere rhetoric exaggeration, a heroic defence of the poor maltreated patient. (We have already responded adversely to some critic's over-response to this weakness.)

We insist on praising Laing for his important achievements. Laing's perception of the interaction between a patient's emotional and mental make-up is novel, significant, and admirable. Whatever else we say of Laing, we also acknowledge with gratitude his factual discovery here, not to mention the shockwave he generated in order to weaken our complacency. Also, we feel, when the shockwave subsides there is left with us the paradox to be resolved. Nevertheless, we expressly oppose his view, regarding the two domains, the emotional and the intellectual: they may indeed interact as strongly as he illustrates, yet we still think they are distinct and can be fairly independent of each other. Rather, we join Szasz who, while viewing them as intellectually distinguishable and indeed two separate entities, yet considers them closely interrelated or interactive. For Szasz the therapist's moral commitment is not only to the patient's suffering, but also to his humanity, which includes his ability to reason and to accept responsibility as an adult member of the community. That is to say, while Szasz observes, following Laing, the patient's suffering clouding his thought, he still insists on the patient's intellectual autonomy.

The greatest difficulty Szasz has faced was to convince his colleagues about the patient's autonomy – his ability to think and to accept responsibility. Even somatic patients are often treated by physicans as not quite responsible adults because of the strain involved and because of the ignorance of the average patient.[6] When it comes to mental patients often the view is that by definition the very option of treating them as adults is too absurd to contemplate. Szasz' claim that the mental patient is a responsible adult seems to run so obviously contrary to so much that has been observed in clinics. This is undoubtedly the case: were clinical observations not so amply confirmatory of the view of psychotics as not autonomous citizens – unable to think coherently and act responsibly – then we would not find the paradoxes of paranoia so paradoxical. The paradoxicality is the very combination of the seeming coherence and the seeming incoherence of the paranoic. Indeed, when an ordinary citizen declares a patient incoherent, he conflicts with expert opinion, yet the very same expert will on other occasions admit a possible incoherence on the part of the same patient. And so, though we cannot dismiss Szasz as easily as his critics do, we can neither endorse his views as they stand, at least not yet.

What we find to be still very important in Szasz' study, as well as in that of Laing, is not so much, or not really, their psychological insights, though they are terrific, especially in their empathy with their patients, and not even their heroic moral attitude (which is in essence nothing but a continuation of Pinel's humanizing the patient as a morally autonomous being), but their sympathetic application of individualistic morality to the sociology of the patient. This is not the place to survey the exciting literature on the sociology of patients and of mental institutions, though we should mention both Lasswell and Rubenstein[7], and Stanton and Schwartz.[8] Rather, we wish to mention Erving Goffman[9] whose remarkable *Asylums* stresses this point over and over, especially the chapter on 'The Moral Career of the Mental Patient'. Anyone familiar with Goffman and even slightly sympathetic to him, will be unable to exhibit the hostility to Laing and to Szasz that is still so common amongst psychiatrists. But, to come back to our point, we are still pressing the psychological aspects of the problem: no one suggests that mental patients are psychologically no different from the rest of the population. And the pressing question is, in which way are they different? Particularly the question is pressing regarding paranoics, and particularly so because of the paradoxes of paranoia.

In other words, the paradox is not a mere logical exercise. We contend that expert clinician-diagnosticians will regularly be incoherent themselves about the question, concerning given paranoic patients, are they coherent or incoherent? It is when considering the patient's ability or inability to accept responsibility that a clinician will observe a patient to be quite incoherent, even if as a diagnostician he has declared the very same patient coherent.

It is possible, of course, that we are here in gross error, that what we observe as empirical observers is not an incoherence of a clinician-diagnostician but a mere play of words, a mere equivocation, a misuse of a homonym: the word 'coherent' means in diagnosis something quite different, it may be claimed, from what it means in clinical practice. Admittedly, to continue this line of thought, a psychiatrist is in an unpleasant position, particularly when acting as an expert witness in court; the same psychiatrist may have to testify about the same patient once as a diagnostician cum clinician, and once as an expert adept in forensic medicine; and he may then use in the same testimony the same word in two quite different meanings. This may regrettably confuse judge and jury (even Szasz' greatest enemies acknowledge the great value of his attack on some evils of forensic psychiatric practices), yet this need not be evidence of any incoherence in the witness' testimony, because it is quite possible to observe in one and the same patient coherence of thought and at one and the same time incoherence of (civic) action. That is to say, this line of thought will lead us to conclude, Dr. Szasz is in error in concluding, if he concludes, from the coherence (and logicality) of the paranoic's thought to the coherence (and responsibilty) of his conduct; whereas if Dr. Szasz admits the incoherence of a patient's thought then his case collapses.

Here, we are afraid, Szasz and we may have to part company. For our own part, Szasz seems to us to be proposing views which we find either unclear or unacceptable. Szasz may be saying that responsibility is not a factual quality but merely a moral quality. We do not accept this: the moral quality is rooted, we think, in the factual ability of an agent to be an agent, to act. (Obligation entails ability.) Or, Szasz may be saying that the mental patient, at least in the early stage of his psychosis so-called – we shall discuss the diverse stages of mental illness later on, so that we may here center on the early stages – is able to act responsibly but is encouraged by the community to jettison all responsibility. The community encourages the patient, says Szasz, to be not responsible; their theory of the madman as not responsible is an incentive to go mad for those who wish to be irresponsible and get away with it; predictions based on this theory are thus self-fulfilling.

Here Szasz and Laing are opposite poles: for Laing the community is run by super-conformists who crush the non-conformist and bring him to an emotional collapse; for Szasz, the community is a bunch of motherly spoilers. What is common to both is the sociological observation that inability to act responsibly is part and parcel of the diagnosis.[10] Now, this makes the sub-clinical or ambulant patient, by definition, not ill at all. This makes illness a function of the severity of the normal standards of a community, the adeptness of the individual to them, and the flexibility with which a community may absorb a deviant as a responsible citizen: a standard functionalist view suggests that stable societies have acceptable functional refuges for psychological deviants where they can function as socially acceptable. Examples are monasteries in various parts of the world, secret societies, academic ivory towers, barracks of mercenary armies, etc. This comes close to Goffman's view of these refuges as total (closed) institutions. Except that Goffman's mental home itself is not functional. A Durkheimian will find the very jail-like character of mental-homes corroboration for the identification of both criminal and mental patient as (seeming) deviants. Thus, Szasz' view of the patient's breakdown as the rejection of responsibility, though highly individualisitc in spirit, is very easy to reconcile with Durkheim's classic collectivism and functionalism. This is so, however, merely because Durkheim's theory of the function of crime is so poor anyhow, that adding psychopathology to sociopathology makes hardly a difference to his theory. All this makes the diagnosis of mental illness too paradoxical to handle, since the fellow may be suffering and classed as a patient by one unspoken criterion and not by another, etc. But this is a mere aside, even though it is no mere thought-experiment: who can say whether modern mental homes are better for psychotics than the mediaeval monasteries?[11]

Now in defence of Szasz who is dismissed by the profession because he views patients as coherent, we have claimed that the same patient may be viewed by the same physician as both coherent and incoherent. The seeming contradiction was then resolved in accord with traditional psychopathology, and it was claimed that a patient may think coherently but act incoherently so that when Szasz views him as coherent he is in error. Szasz, indeed, says that a patient will break down or not depending on incentive; hence, contrary to appearance, Szasz does think that patients can both think and act coherently, and their diagnosis of their own situation reflects societal standards more than anything else. There is, then, a stalemate: whereas both Szasz and his traditionally minded critic may agree that a paranoic patient may think coherently – at least in the early stage of his illness – Szasz but not his opponents will also view his actions (especially his breakdown) as coherent.

The question for the traditionally-minded psychiatrist is, then, can one think coherently but act incoherently? It seems hard to say yes. But, first of all, we can dispose of Szasz' argument thus. Regardless of the incentives or reason or cause which render the paranoic's action incoherent, the fact is that the paranoic is so considered by most clinicians. And so the paranoic just is coherent in thought but not in action. If we could show that it is impossible to be coherent in thought but not in action, then we would have another version of the paradox of paranoia. Now it is logically impossible to have people with coherent thought but incoherent actions. We may even assume (as some philosophers have assumed), that, as far as pure logic is concerned, thought and behaviour are so utterly independent of each other that they never influence each other. If this is assumed to be the case, then our whole conception will have to be drastically revised, both of normalcy and of psychopathology in general, and of paranoia in particular. (Here is another example of Laing's observation that the philosopher's meat may be the patient's poison.) Whatever our philosophy of relations between mind and body, or between thought and action, obviously our diagnostic apparatus takes the interaction of the two to be the normal case. We may, perhaps, still say that for one kind of people there is no interaction between thought and action – to wit, the paranoics. This view is amply empirically refuted by observations of vast areas of paranoic conduct which are as normal as possible; indeed, the diagnosis of paranoia refers to this very fact that in vast areas of human conduct a paranoic is prone to act super-normally. And so, the paradoxicality of paranoia is so strongly expressed in the known fact that even in afflicted areas, within domains of stark delusion, coherence of thought with action is part and parcel of the coherence of paranoia as mentioned in the very definition of paranoia, not to say in its very diagnosis. The paradox thus appears both deep-seated and in the middle of the routine diagnosis.

Let us, then, admit that for every single living person there is some interdependence between thought and action. It will then be most rational for a madman to think about ways to extricate himself from his madness, taking into account, as a starting point, just those junctions of thought and action on which he still is in full control.

Alas! He cannot do so. We do not even know whether he tries, though we have already hinted, we sympathize with Freud's insight that he does, that all mental afflictions are one way or another unsuccessful cases of self-treatment.[12]

Therefore, the paradox stands; to say the patient is coherent here and incoherent there, is, to say the least, puzzling and inviting an explanation, not to say artificial and *ad hoc* and so to be plainly overruled.

No doubt, the paranoic has false premises – so do we all. No doubt, the paranoic may hold different views at different times – so do we all. No doubt, he may switch back and forth systematically – so do we all, incredible though some of us may find this fact.[13] No doubt, the paranoic may have a low level of responsibility – in many societies responsibility is a privilege, in any society anyone may at times feel the burden of responsibility too much even without recourse to a psychotic attack.[14] No doubt, the paranoic may tenaciously hold odd views, even odd integrative principles — so do odd people who are very wise and very sane, so do all sorts of eccentrics and crackpots and cranks who are judged quite normal; so do all sorts of odd societies. The question is, wherein, precisely, lies the paranoic's paranoia?

Now, it is this very question which stops any discussion of the paradoxes of paranoia from being itself a sign of its participant's paranoia. For, were such a discussion in any way leading even to the idea that perhaps, after all, there is no such clinical condition as paranoia, then a major sign of paranoia would thus be revealed. Let us repeat that however much we may agree or disagree with Szasz, he doubtlessly recognized the existence of such a phenomenon, even if not such an entity. And however sincerely we admire Laing, and quite apart from our intellectual dissent from this or that idea of his, we can scarcely brush off with ease all allegations that when he expresses his views the way he does he seems in spots to be a paranoic himself, or at least a persecutionist. This seems so because in certain passage he seems to argue from the paradoxes of paranoia to the outlandish claim that the mentally ill are not ill at all, and even more normal than the so-called normal. In particular this becomes clearer to us when we note that he discusses schizophrenia, not paranoia,[15] and describes its major symptom to be the patient's alienation from his body and his subsequent (and consequent?) general dissociation as secondary signs caused by social pressures.[16] Ergo, psychotics suffer because they are persecuted because they do not conform. Ergo, stop the persecution of non-conformists and you have no appearance of anything like mental illness. Ergo, there is no such clinical condition as psychosis. We confess to find all this odd. We take the paradoxes of paranoia to be in need of resolution and a challenge for the theory of paranoia to solve, not as evidence that there is no such clinical condition as paranoia. Paranoia exists and is challenging us to wonder not only about psychosis; it even makes us wonder what is normalcy[17]; but normalcy[18] still differs from paranoia. The question is, how?

CHAPTER 7

PARANOIA AS A FIXATION OF AN ABSTRACT SYSTEM

We begin with the classic observations of the development of logical thinking from childhood to adolescence due to Bärbel Inhelder and Jean Piaget.[1] We focus on the growth of the individual's ability to consider alternative answers to a given question simultaneously and then test them. We also take for granted for a moment that this is scientific method as described by members of the hypothetico-deductivist school in the philosophy of sciences, especially Sir Karl Popper. Now B. Inhelder and J. Piaget have observed that children perform with varying degrees of success given tasks illustrative of the ability to employ scientific method. Infants think only about one alternative, and one which we can view as the concrete one; adults employ scientific method in ordinary circumstances naturally and with no difficulty, except that when alternatives multiply patience may run out. This gives the false impression that all we need to become scientists is more patience. In fact, however, science deals not with ordinary circumstances but with most unusual facts whose explanations require strong imagination and strong deductive powers.

We shall have to offer corrections to the views just mentioned. We think that Inhelder and Piaget will not greatly resist our corrections. First, Inhelder and Piaget study normal cases and so exclude all cases of fixation and of regression, though these statistically may happen. Second, they ignore the abstract side of research, as we have just mentioned. Before we discuss these corrections we should first briefly explain our ideas of fixation[2] and of regression. These terms come from Freud's classic works and we intend to use them in as strictly Freudian a manner as possible. Yet, not being fully-fledged Freudians we cannot help but modify these terms somewhat. Let us explain how.

Fear reduces ability to examine situations and solve problems. It may lead to a well-known behaviour pattern, that of clinging, like the clinging of infants to their mothers, or when hanging off a cliff. Clinging,[3] no doubt, is regressive, and can become a habit or a conditioning, and we can then speak of it as of a fixation. Freud had three added elements for what he called a fixation. First he had a theory of stages of development, and fixation was within a given stage. We prefer Piaget's stages to Freud's, but do not accept them rigidly either, and feel we can speak of an arrested development even without knowing its course or even without it having a predestined course. Second, Freud saw in a fixation something which feeds on itself, and, especially, in a neurotic fixation a failed attempt at self-treatment which is intensified in an effort to cope with the failure. Since we treat only adult problems in this study, we need not discuss Freud's theory of infants. Yet we accept his idea of neurotic fixation in adults. Third, for Freud all fixations were neurotic because symbolically they lend emotional significance to certain concrete objects. We do not object to his view of neurotic fixations on concrete objects, but we do wish to add to them the clinically observed psychotic fixations on abstract objects.

A word about (intellectual) regression may be in order here, especially in view of N. Cameron's (and others') reported failure to spot any in schizophrenics.[4] Quite apart from the obvious criticism of Cameron's specific study, and quite apart from the possibility of regression, in psychotics or others, from a stage to a lower stage – whether the stages be Freudian or Piagetian – we can be less specific. A patient may visibly regress, in the phenomenological sense[5] rather than in stages according to a given theory, to earlier phases of his or her development. One obvious idea of treatment, of neurotics and of psychotics, is for the therapist to regress with the patient to a stage where his development was still normal and is still intact and take it from there. This idea will accord with both Freudian techniques (not theories, especially not that of catharsis), especially as practised by Marguerite Sechehaye (see Chapter 2 note 26), and with Ey's view. But, not dealing with treatment in the present study, we do not wish to elaborate on this idea here.

Let us apply all this, then, to the findings of Inhelder and Piaget, so as to contrast normal and abnormal developments. Consider fixation, then. The fixation may be on one out of two or more obvious alternatives (ignoring an unobvious alternative is better viewed as an oversight than a fixation), and these may be either a concrete or an abstract alternative or one of a partial degree of abstractness. Now we speak here of fixations in the strict Freudian sense of the word (of a self-reinforcing inadequate treatment); any such fixation may be viewed as a (localised) retardation; it is particularly manifest when it excludes obvious alternatives; it may be transient, or recurrent, or permanent. (We have no intention, however, to compare the onset of the intellectual fixation we discuss here with the onset of an emotional fixation as described by Freud – or by anyone else for that matter; it is only the state of fixation, not its genesis, cause, or mechanism, that we are introducing here.) When we meet an intellectual fixation, we may wish to overcome it by pointing at alternatives overlooked by the subject of the fixation. But its being a fixation makes our manoeuvre unsuccessful: fixation often leads its subject to ignore, at times (unintentionally) wildly ignore, all opposition, disagreement, criticism. Alternatively, fixation allows for the opposition, disagreement and criticism, yet while accommodating all of them *ad hoc* and even with ease. Now this very accommodation of all intellectual difficulty alters the nature of the object of the fixation: the more we overcome such objections the less concrete and the more abstract our fixation becomes. This, then, will be either a misplaced abstractness or a misplaced concreteness or a fixation on an extraordinary object, but one way or another not very relevant to Inhelder and Piaget, and so quite legitimately overlooked by them.

For, and this will lead us to our second correction of Inhelder and Piaget, these two students apply their researches exclusively to concrete matters or situations and not to abstract ones. There are two reasons why concrete cases should suffice for them (our correction is only of the fact that they do not explicitly qualify their theory to concrete and normal occurrences, things, etc.). First, most people find the study of abstract alternatives either incomprehensible or too taxing. Secondly, in abstract thinking errors abound: even trivial errors occur everywhere – even in the highest echelons. This would invalidate Inhelder and Piaget's study were it not confined strictly to the everyday and concrete. For they say that children start with single alternatives and learn in stages to think of many alternatives, whereas, when sufficiently abstract alternatives are concerned, most people stick to only one and most of the rest never master the ability of considering many (see note 25 to Chapter 3).

Well, then, let us go beyond Inhelder and Piaget and consider people who are able to consider fairly abstract alternatives and even to deduce from them less abstract conclusions. Let us envisage a study similar to that of Inhelder and Piaget, but regarding fairly abstract alternatives only, and strictly excluding daily concrete ones. We can even put the above proviso as a criterion for abstractness: the more abstract a statement, the more easy it is to correct it *ad hoc* and even fairly elegantly while admitting concrete criticisms thereof.[6] A concrete theory about observable things can easily be rectified *ad hoc,* but only in a clumsy way. When we see black swans brought from Australia, we have to do something about our view that all swans are white. The easiest is to change our view and declare our previous view not true; but we can rescue it by defining 'swan' to name only white longnecked birds, or we can add a proviso *ad hoc,* such as, 'except Australian swans'. The case of abstract theories is not fundamentally different, but the clumsiness of the *ad hoc* corrections can more easily be masked. The main reason is really not necessarily that abstract theories are more flexible or vague than concrete ones: we have vague concrete theories – about national types, or from astrology – and precise abstract ones, as science has them. But psychologically we are less offended when the words of theoretical science are twisted around *ad hoc* than those of everyday usage, mainly because we are so familiar with established usage we rebel against its arbitrary apologetic change. This can be seen in particular in the case of the apologist who first declares a concept more abstract than it used to be and only then modifies it. This happens to religious readers of the Book of Genesis, to Marxists who are troubled over Marx' economic views, etc.: they do not immediately modify the meanings of terms of refuted theories but first declare their meanings remote from everyday ones. Alternatively, *ad hoc* changes can be introduced in small stages of long periods of time.

The same hold for changes of terms that a person makes *ad hoc* under the force of criticism in order to retain his views of his environment. My concrete view of my environment, when refuted, can hardly be rescued: if the police were really putting my friend in handcuffs it is much harder to insist that he is not a suspect than that he is innocent; likewise it is much harder to insist that he is innocent than that his intentions are pure; much harder to insist that his intentions are pure than that he is a tool of the regime, or more so of God.[7] (Obviously, this is a typical paranoic example; but we can take instances from the history of science: the harder it was to defend phlogistonism *ad hoc,* the more its adherents modified it in a way that rendered it abstract and metaphysical.)[8]

Now, the ability to consider abstract alternatives is largely a matter of intellectual sophistication. And we find it very important to point out that in many cultures no one is capable of breaking away from the accepted intellectual framework and that in no culture is the ability too common. Indeed, we shall soon see, paranoics are people who have this ability but in a limited way. But before coming to this, let us dwell on those who do so with ease. The model of a person who does this regularly is a spy. We mention him first because often he is a rather common man, with little taste for, or habits of, abstract thinking; yet he manages to switch from one set to another, and each set contains both abstract and concrete items. Of course, being little disposed to abstract thinking he tends to take even abstract aspects of the two systems rather concretely, centering more on ritual than on theology in matters religious, as most common folks do, etc. The spy may also have difficulties – intellectual, cultural, even emotional – but he manages to postpone them. Matters differ with social anthropologists who settle down among preliterate peoples, of course: the two systems are seen both abstractly and concretely. Historians of science, or of ideas, or of cultures, tend, on the contrary, to see almost only the abstract and move with relative ease from one abstract system to another.

Again, not all is as simple with intellectuals who move between systems. They may feel the need to stay in the abstract where shifts are easier, and be nervous about concrete applications. Or they may be unable to cast their nets widely. Philosophers who tolerated only Christians, though of different denomination, exist, or those who tolerate only monotheists, etc.

Let us, then, move straight to the paranoic. What is his trouble? In part, at least, there may be a fixation – with some emotional component, that is.[9] The most obvious examples are two; one is the transient psychosis of adolescence which has been observed quite frequently; and the other is the inflexibility of a stranger who may stick to his home system of thinking and behaviour and feelings, at times because he knows no other, or because he is attached to the other, emotionally or otherwise, or because like a European or a traditional Chinese, he deems it much superior, but, and particularly so, at times because he is shaken up and insecure. Of course, a mixture often will occur; the stranger will justify his rigidity by viewing his local system superior, attractive, as well as one to which he is attached and accustomed – and at times he will even confess to a sense of being lost.

The logic of all this is too obvious. To show this let us return to the opposite pole. Let us consider the highly sophisticated case of a person who juggles alternatives and enjoys it. He can be a bi-national citizen spending parts of the year in different countries; he may be our anthropologist moving to the tribe, to the university, and back; he could be our spy, or a double agent; he can be a scientist, social or natural, who juggles systems of thought, or a philosopher, or a poet. A person of this kind, we noted, may feel attached to one alternative, in the sense that he identifies with it. His strong identity may hamper his juggling the other alternatives or it may let him feel ever so free to play with alternative identities. This does not concern us here.[10] He may, alternatively, feel equally at home in all of them (i.e. at home in the world, or in what is common to the alternatives he studies, or at home nowhere – we have instances of these alternatives). This variety of existing possibilities refutes the current neo-Freudian or Eriksonian theory of identity which may be the one propounded by Erikson or not, but which is very popular: every healthy person has a fairly strong and fairly stable ego, which includes a clear stable sense of identity. This theory is empirically refuted, and we may study the cases providing this refutation; but now let us push on. The important aspect common to all these cases is the subject's ability to juggle alternatives, to consider them in some degree of detachment, if not in utter impartiality. At the very least this ability includes knowledge of the existence of alternative abstract systems which may be true, however remotely. This, indeed, is the possibility of theoretical doubt, so essential to any abstract thinking, be it academic or practical, and so naturally considered in our society a necessary ingredient of maturity.

Were the present study a study of all psychopathology, or of mild cases of it, we would center here on the kind of immaturity which academics or intellectuals so often fall prey to. Since they live in abstract systems they can easily show immaturity by preferring one system or another and defending it *ad hoc*. And this procedure leads them to endless petty disputations – petty both intellectually and emotionally. What is mere immaturity in a professor may be fatal in a disturbed person. So let us observe what is required of mature thinking.

The ability we require of a mature person to juggle fairly abstract systems is more than the mere doubt, the mere ability to acknowledge the bare existence of unsuspected abstract alternatives.[11] It is, in addition, the ability to apply some known alternatives to some concrete cases or situations, to employ fairly elementary logic in an effort to use fairly abstract systems as alternative spectacles to see facts and general facts with them – e.g. to participate in political debates with the ability to see the point of this party and that party alternatively.[12] Let us clarify this by presenting the borderline case of a person who knows of the existence of diverse systems but can use only one. He is a person able to perceive alternatives, but to apply to concrete cases, to use as spectacles, only one system. Usually he may somehow learn to use more systems, unless he is not at all interested or unless he takes his ability to use only one system – his – to be proof of its peculiar superiority or even truth. The more sophisticated a person, the more he is able to teach himself to juggle with more than one alternative; and with increasingly abstract ones. And so he will never consider his using of one alternative as a spectacle, his home alternative as it were, to be proof of its superiority in any way, let alone its truth: on the contrary, he will first argue for its superiority and so explain his choice of it as his home alternative.

The borderline case may be an unsophisticated adult, or a young adolescent; or a philosopher or a scientist handling a very difficult problem which he cannot master well, perhaps which he knew only one viable solution to. If he is the latter, then he may grow out of his predicament as soon as he – or someone else – develops a new viable alternative. If he is an adolescent he may grow out of his pet preference in a most natural manner. Yet a very interesting situation often takes place, which we consider as the normal adolescent nearest to the adult paranoic: young adolescents who have known of the existence of alternatives to their native systems all along, but who have no experience of using any of them as spectacles, may feel able at times to change spectacles, but in fits and jerks – not at will, not with ease, and not with the ability to see how one object changes shape with the change of spectacles (very much as some ex-LSD users tune in and out). And, no doubt, various factors are involved in this, from genuine lack of experience to deep seated anxiety.

Psychologists have repeatedly noticed the phenomenon of adolescent fixation or regression. After all Peter Pan stands for this phenomenon in its purely regressive form, and Philip in Somerset Maugham's *Of Human Bondage* for a non-regressive adolescent clinging. What the latter story illustrates is the adolescent's ability to shake off a fixation, though at a cost and in a lengthy process. Now what is perhaps less emphasized but not new is that a passing adolescent fixation can be abstract – indeed they tend to be philosophical. R. D. Laing noted that viewing my own body as a machine is a common one. It is by no means the only one. Indeed, at one place Erikson seems to suggest that adolescent fixations just naturally tend to be abstract.[13]

Now what happens to a youth who does not easily shake off his fixation, whose fixation is abstract, and who is expected to be able to move between different intellectual frameworks?

If, we say, at this stage an abstract fixation takes place, then the result is paranoia. Similarly, a fixation may take place at different times and stages. We shall discuss this in detail soon, but we should immediately conclude from this, first, that paranoia is specifically adult – in adolescents it is not fixed enough to be paranoia proper – and that secondly, in adults it is an adolescent rather than an infantile regression, to use Freudian terminology.

Yet we wish to stress that not all inability to juggle systems, not even all fixations that impair this ability, are psychopathological. Hence, not all such cases are paranoic. Indeed, most of them are not.

A person who has two and only two pairs of spectacles, may be utterly normal, say if he is bi-national and not tri-national, or a member of a bi-party political system etc. Or, alternatively, he may be intellectually limited, say if he belongs to a tri-party system.[14] Thus, if he belongs to a system – political, social, scientific, or psychological – which requires $n+1$ sets of spectacles, whereas he has only n or less, then he has problems; if having n or less sets is a fixation he will need to make a lot of adjustments.[15] Thus we can see, in theory as well as in observed fact, a wide diversity and gradation of fixation. As to the clinical picture of fixation as paranoia, it is not the fixation, it is the emotional counterpart, in brief the subject feeling that he is in strong need of help, which (if it exists) brands him as a paranoic patient proper. Let us avoid discussing the emotional counterpart nonetheless since, we contend, it is a secondary manifestation of primarily a thought-deficiency;[16] let us, rather, present diagrammatically the thought deficiency. We divide all thinking into two: hypothetical (i.e. hypothetico-deductive), when the ability to juggle more than one answer to a given question is present, and absolute otherwise. We also divide thinking into abstract and concrete. We recognized – and this will turn important later on – the fluidity and lack of sharp boundaries here, between the concrete and the abstract. Again, we use the above criterion of the possibility of *ad hoc* amendments: the more abstract claim is given to such amendments with greater ease. The two dichotomies, then, offer four and only four possibilities.

	absolute	hypothetical
abstract	normal adults	exceptional adults such as philosophers, scientists, poets, revolutionaries, binationals
concrete	infants and abnormal adults with fixation (Freudian) or regression (Freudian).	normal adults

The paranoic (and the schizophrenic – see later) has no place in this diagram: he has, in fact, two systems with which he operates. One of these systems is public; he seems to be unaware of it (or he acknowledges its existence as a part of a cosmic conspiracy etc.), yet, he can under supreme pressure, acknowledge its existence[17] (in speech under interrogation, or in desperate action, such as attempting to rescue the lives of dear ones). Also, he is understandably not adept at using it in a normal way since he has little training in that. The other system is private yet he perversely acts on it as if it were public. This, indeed, is crucial. On the one hand there is a fixation here, an inability to shift a framework, to change one's viewpoint however tentatively. On the other hand the acting publicly on a private framework is an expression – in Szasz's sense – of a need for help. And so, whereas the paranoic, like everyone else with more than one framework, operates with one framework at a time, and so consistently, in addition to this, he denies the existence (or the straightforward character) of the public one, and he erroneously uses the private as if it were the public (or the norm for everybody), in a way that will lead the spy to jail and the scientist to the lunatic asylum. Indeed, the lonely scientist in desperate need for recognition may behave as if his novel or outlandish idea is recognized, thereby qualifying as a paranoic (the classic mad scientist.)[18]

Here we may distinguish, quite generally, different levels of sophistication in paranoia. Let us return, for a moment (see end of Chapter 5), to the dualism of nature and convention that permeates Western culture.

Talented youths who discover the conventional nature of their institutions naturally tend to view convention as hypocritical and thus undeserving of any attention of any righteous man. This already leads them to the verge of paranoia, accentuated by the loneliness to which they thus throw themselves. When they mature they learn to live with convention, perhaps at the cost of a measure of self-contempt. There is still the question, does one who knows the difference between nature and convention know either? The paranoic may be surprised to learn that there is the view of the scientific world about nature, which he may endorse. Or he may differ from it – perhaps because it has failed to offer him the expected solace. This divergence from scientific opinion may be self-conscious and so normal, or oblivious of it, or hostile and persecutionist, etc. And so we may have the mad scientist[19] as the most sophisticated of paranoics – having normal views on nature, on convention, on science, yet seeing in the lack of recognition of his contribution, true or alleged, some deep meaning, some conspiracy. This private theory is his downfall.[20]

In conclusion we wish to point out that the paradoxes of paranoia now seem to us completely solved. Each part of paranoia, it seems, is not only known elsewhere but is even a sign of sophistication. Yet the paranoic has both a displacement, and a fixation of it; and neither is normally expected of the sophisticate. The paradoxes then are caused by our ignorance of the place of the spectacles in the patient's mental apparatus (where mental includes both the intellectual and the emotional) and of the background against which he is using them.[21] The logicality of the patient remains unassailed to the end of our analysis.

The strength of the present resolution of the paradox, we feel, lies in a healthy mature skepticism. We do not deny to the end the possibility that the hypothesis of the paranoic is true and the public hypothesis it comes to replace is false. We can also be skeptical about the question who reflects better, is better familiar with and aware of, the public system, the patient or the diagnostician. Whether skepticism is maintained to the last or resolved within science into certitude, high probability, or anything else, all this is of no concern here. We claim, as a part of our theory of the diagnosis of paranoia, that it contains a fixation, i.e. an unexpected element of (seeming) certainty which maturity ought to iron out. And, it may be noted, quite simply, fixation excludes doubt – not so much as a logical licence but as a real ability to have a set alternative to the one on which there is a fixation, or even to conceive of a possibility of the existence of such an alternative.[22] Of course, a paranoic may easily see alternative possible instances of his delusion, e.g. that this or that person present is the enemy, but not to the delusion itself, i.e. the bare existence of the enemy. And if there is even a possibility of an alternative integrative view, then the patient is at that moment not paranoic. Thus, when a patient sincerely conceives of the possibility that he is deluded, he is normal to all purposes and intents, and if we could keep him in that state of mind long enough this would be a tremendous therapeutic success. This is an acknowledged clinical fact which our theory easily explains yet which refutes Freud's, Klein's and others' theories. As to the question, why is it hard to keep the reformed in that mood, it is beside the scope of this essay, and we have no theory to answer it.

It remains to observe that though paranoia is an organization or an integrative system or a systematic delusion, it is nonetheless dissociational even to the degree that some paranoics end up with clearly schizophrenic symptoms. Even when not, however, there is a highly dissociative effect in the very fixation of an integrative system in the obviously wrong place. And here is our criticism of Popper. In Popper's original theory science is a system of conjectures and refutations, perhaps also with certain conjectures surviving attempts to refute them and thus earning for the time being the status of received scientific hypothesis. These may at times be reconciled *ad hoc* with ciples which are not in themselves subject to empirical test yet which act as spectacles, as guiding rules, not less but more than specific scientific hypothesis. These may at times be reconciled *ad hoc* with given tested unrefuted theories (as atomism was with elasticity, see above, p. 43). And from this to the integrative system of the paranoic there is no great distance. This, indeed, is what makes the paradoxes of paronoia of philosophic interest.

We have thus arrived at a proposal for a reform of the definition of paranoia. We find the reform slight in the sense that clinicians are prone to accept it as a matter of course without bothering even to examine it too carefully. We find the reform cogent in the sense that it resolves the paradoxes of paranoia with ease while recognizing their merit and demarcates the sick paranoic, ambulant or incarcerated, from the crazy or mad person who is an eccentric, a crackpot, an honest to goodness scientist, etc. (It even handles well the known borderline cases and suggests new ones, as yet unreported.) We shall later show that the reform has far reaching consequences, both clinical and theoretical.

The reform we propose is this. The paranoic lives in a private world, and speaks in a private language.[23] He treats his own idiosyncratic integrative principle the way most people treat the publicly accepted – institutionalized – one, and is almost totally oblivious of or summarily dismisses the publicly accepted one, at least on points of conflict. A scientist who is identical with a paranoic on all points except this, even one who understandably suffers emotionally from the isolation which this condition incurs – such a scientist will not be classed as paranoic as long as he keeps his sense of proportion on it, namely as long as he remains aware of the privacy of his private views and of the conflict they have with the public views which he is aware of in the normal way. Even if he is feeling persecuted, and rightly or wrongly (and persecution is totally avoidable only by a saint), he need not use this persecution as a private integrative principle, and so he need not be a paranoic. But as soon as this sense of proportion is shaken, as soon as he develops a private world, that scientist becomes a borderline paranoic or worse. It may be diagnostically easy to confuse loneliness with living in a private world; yet only the latter, not the former, is paranoic.[24]

To conclude, we have tried to present a model of paranoia *(vera)* parallel to Freud's model of (sheer) hysteria, with a single desideratum: that the model be immune to the paradoxes of paranoia. We found it an exceedingly suggestive program which lets a lot of known and bothersome facts – bothersome to varying degrees – fall into place, as it were. What is common to both models is the fixation – usually not regressive – which is repressed and is private (here Szasz' distinction between hysteria and paranoia as two different defective sets of communication[25] comes in very handy, though, we feel, it is now superseded) with underlying anxieties and desperate need to both encounter and evade company. What is different is the object of fixation and the consequences from it. The hysteric fixates on a symbolic, emotionally immediate object. He can fixate on an abstract entity too, but on its symbolic[26] – usually phallic[27] – aspect. Thus it will be easier for a hysteric to be, or to be viewed as, regressive. The paranoic will have trouble to relate his fixation to his concrete problems, and he will soon learn to relate every idea to every other idea, well or shoddily. His intellect will disintegrate before his – immature – emotions will. In this his path of deterioration will be so markedly different from the path of a hysteric deterioration – even in cases where the end point is one and the same, i.e. extremely dissociative psychosis or total withdrawal.

Finally, we should notice, even the purest paranoics involve their emotions and hysterics their intellects. But whereas a hysteric can confine, for a while at least, the intellectual damage, so can the paranoic confine, for a while at least, his emotional damage, yet the parallel is incomplete. As a consequence of the paranoic's tendency to compensate intellectually, his emotional make-up is rather immature, though rather relatively intact, yet his intellectual apparatus[28] is damaged and increasingly so. In the case of hysteria, the emotional make-up is both immature and damaged, yet the hysteric can be an intellectual of the first order. But these are the purest – almost pure – cases. Most cases are mixed, especially neurotic-psychotics such as hysterical paranoics.

Whence, then, the tremendous suffering of the paranoic? First, the hysteric suffers too (even when in *belle indifférence,* or apparent discrepancy between the severity of complaints and indifferent facial expression), and shares his troubles with the paranoic. But, unless his case is extreme, the hysteric can be either a person of rather modest aspirations or a person who excels in sublimation, or even both. The paranoic who can likewise sublimate is well-known too, even though we seldom extend him the courtesy of calling him ambulatory or subclinical; he is at times a brilliant intellectual, artist or scientist, before he has a psychotic collapse so-called. Yet, however uncommon paranoia may be, all too often the paranoic is intelligent and so is of high aspirations – at times much too high, due to parental or social pressure – yet his fixation disqualifies him as the intellectual, thinker or artist, that he aspires to be.[29] This difference between hysteria and paranoia is entirely derivative, yet the paranoic's suffering is not therefore less real.

CHAPTER 8

CLINICAL MATTERS

This chapter will be as traditional as possible, though not quite, of course.

The place to begin with, when dealing with clinical matters, is naturally the point where we most strongly differ from standard procedures: we would repeat briefly the alleged possible confusion between paranoia and persecution, and between the paranoic and the schizophrenic (and other) delusions.

First, the confusion between paranoia and persecution. It is true that almost all the delusions of the paranoics are delusions of persecution, but not all of them. A persecutory delusion is neither necessary nor sufficient for paranoia – quite contrary to commonly held views. (A paranoic may have a delusion of grandeur, religious, etc., with no persecution element at all.) The confusion between paranoia and persecution led people to identify both and to interchange paranoia, paranoid, etc., with persecution, being persecuted, over-suspicious, sensitive, etc. Consequently, we know perhaps a lot about persecution, but not about paranoia. It is the permanence, the centrality, the systematization, and the peculiarity of the logicality of the disease, in short, its formal (structural) aspect, that defines paranoia for us. Needless to say, the abundance of the persecution element in paranoia, i.e. the content (emotional and affective) of the disease, is of tremendous importance in the understanding of the patient's *expérience vécue,* his world. The content element is also important for the psychopathology and perhaps even for the understanding of the psychopathogenesis of paranoia. But it has no clinical bearing at all, since it is neither necessary nor sufficient for the clinical diagnosis (of paranoia).

Second, the confusion between paranoic and schizophrenic delusions. The difference lies in the role they play, particularly in the permanence, centrality, systematization, and logic, as opposed to both the semi-systematic quality and semi-logical nature of the schizophrenic (and other) delusions, and to the concomitant existence of Bleuler's fundamental signs of schizophrenia (or the other psychoses).

To confuse matters, we have a third condition, an inbetween psychosis, that Kraepelin called paraphrenia, and that is nowadays better known as paranoid psychosis. It is the case where the delusion is semi-systematic and semi-logical yet where Bleuler's fundamental signs are not found.

TABLE I
Psychoses and psychotic states

		Degree of logicality	
		total	partial
Bleuler's signs	absent	paranoia	paranoid psychosis psychotic episode
	present	schizophrenias affective psychoses	hysteric stupor (hysteric paranoia) psychotic episodes psychoneuroses

All this is important for clinical diagnosis, i.e., for the procedure that demarcates clinically one condition from the others. Therefore, in order to diagnostically differentiate paranoic delusions from those of paranoid psychotics, as well as from those of schizophrenics (and others) two qualifications should be considered.

One qualification concerns the very nature of the paranoic's delusion, its formal aspect just mentioned: its systematization and good logic. In that, paranoia stands in a category apart. In contrast to the systematization and logic of the paranoic's delusion, we have a delusion where its formal structure is characterized by semi-systematization and a less perfect logic, 'semi-logical', as the term goes. Such a delusion will not be called paranoia at all but paranoid (id = similar to; hence, paranoid = similar to paranoia, but not identical with it). In technical terms we speak of a person exhibiting a semi-systematic and semi-logical delusion, as suffering from, or being in, a paranoid state. Here comes the second qualification for the differential diagnosis of paranoia from other psychiatric conditions. This qualification is, the exclusiveness of the delusional state as a pathological (morbid) sign in the overall (total) clinical picture. Here paranoia is a condition characterized by the fact that in addition to the particular systematic delusion (itself, as mentioned, differentiated from the paranoid semi-logical delusion), there are no other pathological (morbid) signs to be found, there is no disturbance in any sphere of mental life, "... in thinking, affectivity, volition ...", etc. It is interesting to note, using our two qualifications, that there is a clinical condition, similar to paranoia, in that except for the delusion, there are no other pathological signs present, yet the delusion is a semi-logical (hence paranoid) one; this condition is called, not paranoia, but, as already mentioned, paraphrenia, or paranoid psychosis.

Both qualifications are illustrated in the table of psychoses and psychotic conditions above. It is here that we shall have to put the differential diagnosis of paranoia, of schizophrenia, the affective psychoses, and the organic, both the psychiatric and the neurological diseases. For schizophrenia we shall look for its fundamental signs, namely, the disturbance of association (the process of thinking in general, as Bleuler already understood it, rather than of association[1] proper), the disturbance of affect, the ambivalence and autism.[2] Now, a schizophrenic is so diagnosed if the fundamental signs are found; contrary to the popular view, there is no necessity at all for a schizophrenic to have a delusion. Delusions, hallucinations, catatonic manifestations etc., are all accessory, not fundamental signs, in schizophrenia. When a delusion is accompanied by fundamental signs, the diagnosis is of paranoid schizophrenia.

The same line of clinical reasoning, i.e., looking for additional symptoms and signs, except for the delusion, is applied to affective psychosis and organic diseases, etc. It is thus that we can find a paranoid state as part of manic and of depressive conditions, and as part of a puerperal (postpartum) psychosis, senile dementia, general paralysis of the insane, Parkinson's disease, etc. In summary, then, *paranoia vera* will differ from other psychiatric conditions in two ways. It will vary from all paranoid states by the fact that in all paranoid states the patient may move his fixation, though with difficulty, form one integrative principle to another; a paranoic does not. And in similarity (symmetry) with the paranoid psychosis, in paranoia there are no pathological signs, except for the delusion.

In these few paragraphs, we hope we have summed up in a rather abstract as well as summary manner, much of the accepted preliminary critical material. We hope that the systematic theoretical bias is clear in them but we cannot here go into detailed description except of the case of *paranoia vera,* which we shall now present. We shall return to all this in the next and final chapter, where our theoretical bias will take over, and where we shall question much of the clinical material mentioned thus far in this chapter. We prefer to discuss the accumulated clinical data of paranoia first, as we deem these more accurate than, say, those of schizophrenia.[3]

Let us now dwell on empirical matters as much as necessary for the achievement of a clear distinction between *paranoia vera* and paranoid schizophrenia. Since we claim that fundamentally these two are diagnostically more or less the same, we have to discuss clinical descriptions and diagnoses, as well as some clinical courses, going from the more factual, where the difference is perceived, to the more theoretical, where (we claim) it diminishes in size.

Well, then. The pure paranoic is by definition not a disturbed person in the realms of his thinking capacity, affect, and actions. Indeed, the first question about him is, how come he has met the psychiatrist in the first place. In contrast, the paranoid schizophrenic, as any schizophrenic, displays Bleuler's fundamental signs such as the disturbances in the realms of thinking capacity (mainly looseness of associations, discontinuity in the flow of thinking which is seemingly more or less wild association), in affect (flat or poor emotional life – we tend to question this – discrepancy between thought and emotion, and seemingly inappropriate gesture), as well as in action (the discordance between intention and performance; ambivalence), not to mention autism.

So much for the immediate difference between paranoics and schizophrenics of all sorts. Now a schizophrenic, i.e. one who exhibits all these signs, may be of the simplex type; or he may be hebephrenic, catatonic, or paranoid; or he may be a combination of all these. To be paranoid he has to experience delusions and/or hallucinations.

So much for signs. As for diagnosis, paranoia displays what we described when we discussed the paradoxes of paranoia. Paranoid schizophrenia displays delusions which may, but need not be, highly sophisticated.

So much for difference in signs and diagnosis. As to course, paranoia is today deemed chronic, and some clinicians claim that every paranoic (unless he dies young) ends as a chronic schizophrenic.

Paranoid schizophrenics are either chronic, sometimes ending demented, or get cured, with or without relapses. In the last case, a post-psychotic personality-defect may be expected.

We have mentioned the curious variant possible here, which has been recorded, which is in between paranoia and schizophrenia, called by Kraepelin paraphrenia and today known as paranoid psychosis. It is characterized by two features. The one is a delusional system resembling that of paranoid schizophrenia in that it is less coherent than that of paranoia. Yet, it shares its other characteristic with paranoia in that in it thinking, affect, and action, are fairly well intact; there are none of the fundamental signs of schizophrenia. The course of this case resembles that of paranoid schizophrenia.

We have repeatedly left open the question, how come the paranoic meets the psychiatrist in the first place. Let us look at this for a while, as well as at the symptoms and signs of paranoia as presented by a concrete patient.

The symptoms of paranoia, namely the patient's complaints, are either brought to the attention of the clinician by the patient himself or by a third party – police, neighbours, injured bystanders, etc. In the latter case, we may assume that it is the patient's decision that causes the third party to bring him to medical attention. He may do so by a direct action, for example, by asking police protection or an act through which he becomes a public nuisance. Alternatively, he may do so as an unintended consequence of his action. He may force people to bring him to medical attention merely by accident, e.g. when his assessment of his nuisance value is by chance too low, when he wanted attention but by overacting received too much of it. The difference between the case where the patient attracts medical attention directly, and the case where he does so indirectly, may be very small indeed, since he may ask for a doctor's protection just as much as for police protection, etc. What is common, then, to both these cases, that is to say, what symptom a paranoic may exhibit, is never his confessed knowledge of his peculiar mental condition as mental, though, of course, he is all too well aware of the peculiarity of his condition one way or another. Clinically, then, the symptom of paranoia is never manifest, but always a by-product, often even accompanied by another such symptom, namely the strong denial of any major symptom and any mental illness whatsoever.

As for the observed signs of paranoia, the above reported signs are of course the major ones. As to other signs, to make things even more sophisticated, it is their stark absence, which is the second sign that traditionally draws the clinician's attention to the possibility of a diagnosis as paranoia. To be more detailed, we could note that the patient is usually examined for his thinking, affect, will, personality, orientation, perception, the mental faculties of memory, remembering, attention, concentration – and to the clinician's mounting surprise is found not wanting[4] (no matter how expert, and how quick to suspect a case of paranoia, the clinician is repeatedly surprised at the presence of a pure case of paranoia, and dismisses the impure case as no case at all; on which more later). In particular, the patient exhibits no hallucinations[5] and no bizarre behaviour, his activities are easily explained by him in a cogent manner by his single delusion. Indeed, he is often so markedly active in a manner consistent with his delusion that French clinicians traditionally refer to him as *persécuté-persécuteur:* his attempts to escape his delusional persecutor or to fight him back are so marked that even a non-clinician may suspect that he wishes to keep his persecutor alert.

The signs of paranoia *(vera)* thus far, then, are three: first, the patient shows only incidental symptoms while denying his being mentally ill altogether (as we all do); second, he is fairly normal otherwise (as hopefully we all are); third, he has an *idée fixe* (as many of us have). Our diagnosis, then, thus far excludes many cases of mental illness, but includes many normal cases as answering the description of paranoia! We must, then, say something about the diagnosis of the paranoic's *idée fixe* itself.[6]

The paranoic's *idée fixe* is, of course, a systematic chronic delusion. Its being systematic is a matter of pure logic: the patient is more, not less, logical than the average person, or even the average *idée fixe* bearer. Its being chronic is, of course, no easy matter for diagnosis, and so all diagnoses of paranoia may be mistaken and should be of paranoid psychosis or even of paranoid schizophrenia, just as a diagnosis of paranoid psychosis may turn out to be mistaken and the case may be of *paranoia vera* (accompanied by a merely temporary confusion) or of paranoid schizophrenia (accompanied by permanent dissociation). (Of course a case of paranoid schizophrenia may hardly ever turn out to be only paranoic, but paranoia leading to dissociation should in retrospect be reclassed – perhaps only after decades of hospitalization have passed.) So much for the systematic and chronic character of the delusion. As to the delusion itself, its primary axiom (or premise) is self-reinforcing or self-verifying (what Popper calls a reinforced dogmatism):[7] assuming the primary axiom (or the premise) to be true, the clinician sees, in addition to the distortion exhibited by the patient, the explanation of all difficulties which the patient may encounter; he then may easily run the gauntlet together with his patient. (It is intellectually easy, that is; emotionally it is never easy to run the psychotic's gauntlet.)

It is on this very point that we find our having chosen *paranoia vera* particularly fortunate, for the more dissociated the patient, the harder it is to predict his next move, and the more integrated the paranoia, the shorter it takes the psychiatrist to learn to predict the patient's very next move, and with a surprisingly high degree of precision and success. It is this high predictability which is, as ever, the standard high sign of any fixation. There is no doubt that some dissociation, being universal, is to be found also in *paranoia vera* and so successful prediction is never fully successful. Nevertheless, even though a dissociated move is hardly predictable, the onset and increase of dissociation may be. Of course, for the time being, this is seldom attempted.

So much for the symptoms and signs of paranoia. This concludes the traditional matter of clinical diagnosis, and we still find it wanting. To conclude the matter of diagnosis here, we claim, will make much that is normal, even though dogmatic or fanatic, as paranoic. To avoid this, we think, the emotional components of the patient must be considered as well. The patient's expression of his suffering may be the description of a persecution mania well in accord with his *idée fixe,* or it may be expressed otherwise. One way or another, we contend the diagnostician will observe with ease an emotional parallel to the logical impasse in which the patient is stuck: whereas the dogmatic may enjoy the finality of his logical impasse, whereas the fanatic may translate it into a frank and unproblematic and even self-righteous aggression, the paranoic is cornered[8] and feels cornered[9] and even desperate. This concludes our diagnostic description of the case.

A corollary to this addendum is that the paranoic shows no primary but many secondary symptoms of anxieties and depression (not in Jackson's sense but in the sense of a secondary reaction, or reactive depression) – acute, sub-acute or chronic. This corollary agrees with known clinical studies, especially the reported peculiar experience of paranoics known as *expérience vécue, le vécu délirant,* etc.

The signs of the disease are usually observed in the third and fourth decades of life.[10] It is reported that women are statistically more frequent victims of the disease.[11]

The course of the disease is described in traditional texts in two opposite ways,[12] as noted already by Ey.[13] We shall attempt to reconcile them here. On the one hand the literature describes a process of increased integration.[14] The *idée fixe* which begins as relatively localised or limited, naturally becomes so vast that hardly anything can catch the patient's attention without his interpreting it as deeply related to his *idée fixe.* We say 'naturally' because, as we have noted already, normal people, and animals too, under stress or in a state of any concentrated attention, whether in an attempt to relieve the pressure or in order to solve any other problem, tend to see everything in relation to the problem at hand; this is centrism[15] in Piaget's terminology, and Gestaltists have more terms for it. Konrad Lorenz noted already in 1935 that fear stupefies. "One source of error that is overlooked in many maze and puzzle-box experiments is that panic, particularly in the highest animals, reduces the mental faculties to a minimum. If a mentally high-ranking and therefore easily excitable bird is even slightly frightened in the course of a detour experiment, his mental performance immediately falls far below that of a much duller animal, which does not react with fear to the same environmental change."[16] It may be interesting to note that usually we consider with great ambivalence the total suppression of fear and see it partly divine and partly robotic or even spider-like or insect-like in a manner exploited in many a horror movie. This, naturally, explains Laing's phenomenon of the patient's sense of his body as a robot: he feels the weight of his problems and fears and can only break away from them by enormous emotional effort. This effort, of course, is self-defeating as the patient's centrism is not thereby relieved. The centrism[17] of the paranoic, to repeat, is quite normal, though its weight and duration are not. This centrism, naturally, and quite logically, leads the patient to egocentrism[18] and to megalomania:[19] since his *idée fixe* somehow relates to him and somehow gains in import and centrality, by becoming central it makes him likewise central, and with an inexorable logic. On the one hand, then, the onset of the disease is increasingly integrative; on the other hand, it is often reported, and we corroborate it, dissociation may also set in, and in diverse forms, up to and including schizophrenic breakdowns.

The explanation of both trends – integrative as well as dissociative – will easily avail itself once we apply here Freud's insight, already mentioned before, of all disturbances as faulty self-treatments which reinforce themselves in vicious circles. Particularly where fixation of any kind is employed, we may expect the patient, on this Freudian theory, to increase the dose of self-treatment with the increase of suffering and thus cause further increase of suffering. It is our view that paranoia is a cause of suffering in which the patient is trapped, and so, applying Freud's insight we conclude that the treatment i.e. the systematic logical trap, will increase, whereas the suffering will increase as a result; we conclude then, that this increased suffering is the main factor behind the dissociation. The dissociation itself, then, from paranoia on to schizophrenia, is quite derivative[20] and so need not always take place,[21] this in accord with known observation.

So much for symptoms, signs, onset, and course of the disease. We see no need to go into etiology, but in deference to tradition let us report earlier views. Kraepelin (and Bleuler) suspected physical brain damage of some kind or another on general mechanistic grounds. Freud declared all mental illness to be psychogenic, though in the last resort he was a mechanist too.

It is very interesting to observe that Freud was concerned to explain the persecution of the paranoic by normal psychoanalytic processes of projection and internalization.[22] Of course, we reject these as both inadequate (explaining too little) and false (as the fixation is abstract not concrete). Freud had nothing to say about the *idée fixe,* perhaps because he saw it as no different from other fixations, hysterical or obsessional.

As an aside we may also observe that Freud considered both paranoid schizophrenia and paranoia to be latent homosexuality and troublesome undischarged cathexis or mental energy of a narcissist-homosexual type.[23] This should have made him confuse the two diseases. At one time he did; he then relented despite this theory and he relied on nothing more than on strong common sense. We find it important that Freud stressed the patient's suffering and regret that he overlooked the patient's peculiar thought-mechanism, or rather, felt it to be no more than a peculiar case of projection-internalization mechanism. Already Melanie Klein rejected Freud's implicit idea that all (normal and pathological) thinking is by association, projection, and internalization; she suggested that paranoia is a specific thought-impairment; indeed, a thought impairment which is a regression. Anyway, even Freud's analysis of the disease as latent homosexuality[24] has been overthrown by his own disciples, or at least put into severe doubt: at the very least we should not be blind to non-paranoic latent homosexuals, not to speak of non-homosexual paranoics (or, to be a pedantic Freudian, no more than normally latent homosexual paranoics) etc. Be it as it may, for our part all this is a matter of the past since we explain paranoia not as an emotional disturbance but as an intellectual disturbance which has a strong emotional bearing as we have described: we think that all manifest emotional symptoms, strong as they are, are merely derivative. True, unless the patient suffers he is no patient; but unless his impairment is intellectual he is no paranoic.

We have concluded our restatement of a diagnosis and of etiology. Before conclusion we wish to notice studies of the intelligence and thought-patterns of the patient. This naturally takes us entirely away from Freud to Piaget. We must note, however, that the distinction between the intellectual and the emotional is purely our intellectual distinction, not a mark of any sharp separation. And so it is not surprising to see a study, by Gouin-Décarie who is both Freudian and Piagetian, on the parallel development of both affectivity and intelligence in the young child. On the contrary, we find it surprising that only one such study is available thus far. In any case, there is nothing in such studies that requires further discussion on our part. The intelligence tests of the paranoic show no disturbance, and at times high quality. From this, however, we should not conclude the same of his thinking: intelligence and thinking are distinct. We think, in particular, that in paranoia there is a disturbance – not of intelligence but of thinking. The peculiarity of the thinking disturbance is precisely its clinical elusiveness noted before in this chapter as well as in previous ones, when discussing the paradoxes of paranoia.[25] Oddly, this idea is novel only when applied to paranoia; when applied to schizophrenia it is well-known: it is a commonplace that the schizophrenic's intelligence is normal whereas his thinking is severely impaired. What we suggest is that the thinking of the paranoic is impaired in a more sophisticated, higher level, more abstract manner, than that of the schizophrenic.[26]

And now to our own conclusion. We find the present chapter rather superficial, though we have tried not to omit anything of substance from the clinical literature that has come our way, whether books, scientific periodicals devoted to the topic, or genuine clinical experiences.[27]

What particularly disappoints us in all this is, of course, the enormous insensitivity to the paradoxes of paranoia and all that they entail. In particular, even the very existence of an integrative principle in paranoia is something the literature treats as no more than a mere sign; indeed, even the label 'integrative principle' or its cognate is not to be found in the literature even though already Kraepelin did describe it as a sign, when he talked about 'deep-seated change in a person's whole outlook' due to his delusion. Whether Kraepelin considered this a mere sign or a key factor we cannot tell.

Now, what we wish to do next is, first and foremost, to offer as a diagnostic principle of paranoia an abstract fixation that becomes a private integrative principle in a self-compensatory move.[28] Second, we wish to characterize the dissociation that such a mental illness may undergo.

COGNITIVE IMPAIRMENT IN SCHIZOPHRENIA

It is common knowledge that a severe cognitive impairment in schizophrenia exists:[1] it is part and parcel of Bleuler's very diagnostic definition of schizophrenia.[2] It is likewise common knowledge that this impairment has a specific form; as the impaired mode of thinking is usually characterzed as a primitive and archaic form.[3] Evidence supporting this is amply supplied from three independent sources: from anthropology, from child psychology, and from psychoanalysis.[4] The literature regularly, even if vulgarly, contrasts adult-maturc-logical-occidental man with childlike-immature-mythical-tribal man. The adult is, of course, also contrasted with the child proper and likewise with the neurotic. Lévy-Bruhl,[5] Piaget,[6] and Freud,[7] each in his turn, described roughly the same strange picture of a special strange world, where wish always has priority over reality, where logical contradiction or causal impossibility is not always forbidden, where concrete thinking governs space, time, causality, people, and events. Prelogic and paralogic, regressive and autistic, primitive and archaic forms of thinking, they all and equally tyranically dominate the child, the savage and the neurotic; and now, ever since Bleuler, it is common knowledge that this very form of thinking is also characteristic of the schizophrenic.[8] So much for the common knowledge.

Now to begin with, we do not claim to be clear enough about all this, even if we are the only ones who are so slow to grasp the concepts involved. Of course, we agree that in psychosis thinking is impaired, one way or another, and, indeed, even in neurosis. Hence, it is not the general idea that we fail to grasp – and incidentally endorse – but the concept of the mature logical and adult as opposed to the child-like and primitive.

Well, then, what we want to dispute, in brief, is the very loose use (indeed, the abuse) of the label of primitive thinking; it was used too facilely, carelessly, and therefore masked very important and real questions, such as that of the demarcation between a dreamer and a child, between a neurosis and a psychosis, between primitive and cultured, not to mention between a child's and a primitive's dream. etc.

To begin with dreamers and children. According to psychoanalysis both the dreamer and the child display a primitive form of thinking. But this, of course, very insightful as it was, is pathetically inadequate, when taken too seriously or literally. For, after all, it puts all children's attitudes on a par with that of a dreamer, whereas a child can dream at one time and display realism at another; and a dream, a daydream in particular, can be the merest fantasy and it can be the germ of a realistic plan. We cannot seriously put all these in one kettle of fish. And the same, of course, holds for the other distinctions, between the primitive and the cultured, between the sane and the neurotic.

Similarly, we find so very important both what is common to and what differentiates between dreams and games. Children do not always succeed in fullfilling their wishes in true life situations, so there is always the game or play-situation, where primitive forms of imaginative wishful thinking take place; so is dreaming; yet, if dreaming – why playing? And vice versa.

In order to put things more sharply, let us consider the child's own view of his own dreams:[9] one sees immediately that primitive forms of thought are reflected in the dream itself; also that the dream itself is considered in a wakeful yet primitive and archaical manner: infants often say – till the age of 7 at times – that what was dreamt took place in a manner permitting everybody to see. Yet differentiation of sorts comes very early: even while describing a dream as if it were a public event at times the child knows dreams from reality. After the age of 7 anyone will conceive dreams quite differently from public events. It is as if the post-infancy dream is a partial or local form of primitive thinking (regression?), while the infant's form of thinking is on a general quasi-global level (and up to the age of 3 it is truly global).

We find it a bit hard to employ the contrast between the infant and the post-infant modes of dreaming to the primitive and the cultured respectively, and even to the sane and the insane respectively, except that we do agree on one point. Before the age of seven, day-dreams are very much like delusions in the sole sense of lacking any reality test, but decidedly not in the sense of any pathology (for example, while displaying no signs of underlying anxiety). Moreover, as we argue (see p. 69), a psychotic under stress may temporarily force himself, though suffering great pains of severe anxiety, to apply sane reality tests; whereas infants under stress simply function less well than normally, and in all respects.

The same criticism should apply to the distinction between neurosis and psychosis. Neurosis tends to be a local regression, psychosis a global one. The cognitive impairment in schizophrenia is not in that it is a primitive form of thinking, but in that it is a low level integrative principle, and a global one at that.

Our demarcation of paranoia as a psychosis, lies in that we consider it to be a global regression (or fixation) of the intellectual apparatus, yet (in contradistinction to schizophrenia) not a regression to a concrete level of behaviour, but to an abstract one. Nevertheless this abstract level is still a lower one in the hierarchy of abstract behaviors.

Moreover, we wish to stress the intention: however local a neurosis – symptom neurosis, to use Allen Wheelis'[10] term – is intended to be, already Freud's early studies in hysteria show how a crack in a person's character tends to spread, and finally, for example, prevents a patient from swallowing anything whatever.[11] This, we would say, is the outgrowth of neurosis in the direction of psychosis (and will vindicate, again, Freud's claim that the two are essentially identical, by presenting it as a good approximation).

Now, we have no intention of further criticizing the vagueness of the contrasts we have presented here (in the first paragraph of this Appendix). The sensitive reader will not be surprised to learn that the fuzziness of the contrast permitted the inclusion of all sorts of impaired thinking as primitive. The most incredible example is that of aphasia,[12] i.e. impaired thinking due to brain damage. This, of course, is a regression of psychiatry to crude 19th century materialism.[13] So we shall not go into detail in the direction of listing the diverse forms of impaired thinking mislabelled as primitive. Rather, let us observe recent attempts to make the distinctions at hand less fuzzy than they used to be.

The ideas criticized here as fuzzy were particularly popular up to the thirties. They still are, but meanwhile the *avant-garde* came with finer distinctions and with proper tools to observe them. Indeed, the refinement came by studying other cognitive factors which were sought and observed, as well as purely perceptual ones which were observed too.[14] Today, it seems, there is a tendency to observe some disintegration in psychotic patients manifest in both the cognitive and perceptual failings. This already suggests an idea about a normal person's integrative principle impaired in, or during, psychosis. This integrative principle is usually a structure, linguistic or holistic-personal, or both.

Need one say that this trend is hardest hit by the paradox of paranoia?

Let us, however, briefly describe at least the most important trend, i.e. the one which brought about most of our knowledge about the impairment of abstract thinking in schizophrenia, so as to provide the reader with its 'testological' flavour: we have said enough by the way of criticism.

The psychiatric tradition traces the beginnings of intentional, organized, and cumulative study, of thinking in schizophrenia, – in the development of the special psychological testing methods, due to L. S. Vigotsky in Moscow in the thirties and to K. Goldstein in Berlin then and in the forties in the U.S.

Vigotsky[15] was impressed by the findings of J. Piaget in the psychology of thinking in the child, and used these findings in his elaboration of a specific test for concept formation. Regarding the theory of the origins and development of concept formation, Vigotsky held a view opposite to Piaget's; yet as far as the test itself is concerned, this divergence does not matter. Vigotsky has fathered a whole group or kind of tests. Let us, however, describe the one he himself has devised. The material for Vigotsky's test consisted of 2 pieces of wood, varying in colour (5 different colours), form (6 different forms), thickness (2 different thicknesses), and size (2 different sizes, one big and one small). The subject has to classify these pieces of wood in four groups. Indeed, there is only one possibility of conceptual classification. The definite groups are as follows: big thins (6 pieces), small thins (6 pieces), big thick (5 pieces), small thick (5 pieces). This is, obviously, the one and only combination of two available criteria – size and thickness – which permits a simple classification into four groups: one has to take into account that the criteria of form and of colour are distributed in such a way that they cannot smoothly follow the classification just mentioned.

The aim of the test is to appreciate by what successive steps the subject comes to conceive which are the helpful (useful) criteria. In order to help him, the pieces of each group carry, on the facet facing the table, a meaningless monosyllabic word. When the subject has proposed an inadequate classification, the examiner turns two pieces in a way to show that the group is incorrect since the syllables of the two pieces are not identical. The test is conducted in two stages. In the first stage, the small red thick triangle is taken, turned upside down and it is shown to the subject that the piece belongs to the group 'mur'. The subject is then asked to put together with this piece all those pieces which, he thinks, enter this group. If, for example, he utilizes the criteria of colour, he will put together all the red pieces. Then the examiner will turn upside down the big square thick red pieces (for example) and show the subject the syllable 'lag'. The subject will thus see that the criterion colour is inadequate. The test is continued in this manner. The presentation of the test is arranged in such a way, that after 3 corrections, the subject has had before his eyes an example of each of the group. When the subject succeeds in forming a group, he passes to the next.

The second stage of the test examines articulation: the subject has to describe the principles of his classifications, more specifically he is asked to answer the successive following questions, referring to: a. principle: how are the pieces being classified? b. definition: describe the pieces of this group in relation to the other groups. c. common characteristics: what have all the pieces of a group in common? and d. differences: in what aspect does a (particular) group differ from another (particular) group? The scoring is a very complex one. It is based upon the subject's behaviour in relation to three aspects of the test: interpretation of the task (how the subject interprets [understands] what he is asked to do); trials (and errors) of the solution; and discovery and the understanding of the correct solution. For each one of the aspects, later on, two famous American followers of Vigotsky, Hanfmann and Kasanin, have added a characterization of the test in 3 levels, the conceptual level, the intermediate level, and the primitive level. So much for a description of the Vigotsky test which, by and large, was found to be too difficult to administer. It is also very difficult to pass: to fully succeed in it, the examinee has not only to be of a high degree of intelligence, but also to have accomplished at least secondary education. Nevertheless, practical considerations aside, Vigotsky's work gave a very important impetus to the study of thinking in schizophrenia.[16]

The other important group of special tests of conceptual thinking, was designed by K. Goldstein and described in detail by Goldstein and M. Scheerer in 1941, in a classic monograph on *Abstract and Concrete Behaviour*.[17] Let us briefly describe their so-called 'object sorting test'. The purpose of the test is 'to determine whether the subject is able to sort a variety of simultaneously presented objects according to general concepts; to shift these frames of reference volitionally' (p. 81). The objects are (the list is here abbreviated):

Article	*Material*	*Colour*	*Form*
Toy spoon	Metal	Silver	Oblong
Toy noise maker	Wood	Brown	Oblong
Toy hammer	Metal and wood	Silver & brown	Oblong
Toy dog	Porcelain	White	Small
Chocolate cigar	Food	Brown	Cylinder
Ball	Rubber	Red	Round
Pipe	Wood	Brown	Curved
Cigar	Tobacco	Brown	Cylinder

etc. (27 objects for men; 29 objects for women).

The subject is asked to group articles with an object which he has selected himself; then to group articles with one selected by the examiners (handing over) . . . In the second stage of the test, the subject is asked to group all the articles which he thinks belong together (sorting) . . . At last the subject is asked to arrange all articles in still another way. If he does not comply satisfactorily, he is presented with new groupings by the examiner (shifting). Obviously, the articles can be grouped as to use, situation, colour, form, double occurrence in pairs, and material (p. 82).

In every stage of the test the subject is asked to explain why he grouped the articles as he did, etc.

The authors explain. There are two qualitatively different attitudes, concrete and abstract. Whereas normal people can shift from concrete to abstract concepts and back, depending on circumstances in a fairly obvious – and normal – way, schizophrenics (they do not discuss other psychotics) systematically fail to engage abstract concepts. The same applies to aphasic patients. Indeed, Goldstein's work, now followed by Norman Geschwind[18], is chiefly neurological and centers on aphasia. His work on schizophrenia is a mere branching off of the assumption that both aphasic and schizophrenic thinking is impaired. It is hardly necessary to say that Goldstein and Scheerer on schizophrenia as non-abstract has been corrected – chiefly by Chapman and Chapman[19] – that is to say, not that the abstract is missing in schizophrenic thinking but that it is concretized.[20]

For our part, we hardly disagree, though we prefer fixated to concretized, but rather, we wish to emphasize that any theory of impairment of thinking that is not sensitive to the hypothetical nature of all human thinking is pre-Einsteinian, and is hit by the paradox of paranoia.

One last mention of diverse tests.[21] The Rorschach test, where amorphous blots arouse associations in patients, the TAT (Thematic Apperception Test) where the patient is asked to narrate a story when observing a suitable picture, cartoon, etc., sentence-completion tests, even mathematical completion tests, IQ tests, the Wechsler-Bender test akin to the IQ test, and so on and so forth, were all used one way or another to diagnose patients, to decide whether to release patients from institutions, etc., and even as to whether psychotherapy is advisable. We find little or nothing to say on all these.

In conclusion, we wish to make one general observation on the literature surveyed in this appendix. It is quite remarkable that practically all attempts at an inter-disciplinary look at psychosis, often also neurosis, not to say psychopathology in everyday life, usually rest on, or come up with, or suggest the profound insight of, nineteenth century evolutionism. The reader, who is still hopeful that this idea has some promise is strongly advised to read in conjunction K. R. Popper, *The Poverty of Historicism*[22] and P. B. Medawar, *The Future of Man*.[23]

APPENDIX II

FREUD'S VIEW OF NEUROSIS AND PSYCHOSIS

It is common knowledge that Freud's main contribution to psychopathology is not so much the point of demarcation as the point of similarity. Thus, according to Freud, everybody is a bit of a neurotic and a bit of a psychotic. This, however, need not make psychosis only one extreme as opposed to health, with neurosis in between the two. Yet, Freud insisted that this is so. Thus the most obvious and clear symptom of psychosis, says Freud in his posthumous *An Outline of Psycho-analysis,*[1] is that of schizophrenia, of a 'psychical split' or a 'splitting of the ego' (Chapter 8, p. 202); yet this very phenomenon is equally common in neuroses, for example in fetishism.

There is no doubt that this was a great discovery, and that our attempt to demarcate neuroses from psychoses as matters of kind rather than of degree should not obscure Freud's discovery. Moreover, his example, fetishism, is extremely intriguing, since a fetish can be a symbol and thus neurotic, or an abstraction and thus psychotic; it is, we propose, an ambiguous borderline case. Yet, we repeatedly claim, the very existence of borderline cases, (as distinct entities) and of mixtures, forces us to reject the idea of psychoses as merely extremely severe neuroses.

Now Freud talks specifically about neuroses and psychoses in three distinct places: 'Neurosis and Psychosis', (1924);[2] 'The Loss of Reality in Neurosis and Psychosis', (1924);[3] and 'Fetishism', (1927).[4] Let us notice these. 'Neurosis and Psychosis' (1924)[5] takes as its starting point a hypothesis 'raised in other quarters'. The editors of the Standard Edition explicate in the first page (S.E. 1961, p. 149) a hint to Goethe, but not to the 'other quarters' which, we suggest, is Kraepelin. The hypothesis that Freud has to offer, he says, he developed 'in connection with a train of thought raised in other quarters' – but he does not even say which train of thought, except that it 'was concerned with the origin and prevention of the psychoses'; the definite article in 'the psychoses' is a slip of Freud's pen: 'psychoses' with no definite article is required here. Perhaps a slip of Freud's pen should not be let go unanalyzed, but we do not know how to analyze it.

'In connection with a train of thought raised in other quarters, which was concerned with the origin and prevention of the psychoses,' then, 'a simple formula has now occurred to me which deals with what is perhaps the most important genetic difference between a neurosis and a psychosis' – so here we have a genetic demarcation between the two; we quote it in full, with Freud's own emphasis: *'neurosis is the result of a conflict between the ego and its id, whereas psychosis is the analogous outcome of a similar disturbance in the relations between the ego and the external world.'*

We notice, first, an uneasy formulation. Why not, conflict between ego and id causes neurosis, between ego and external world causes psychosis? The answer may be that the external world need not be purposive and so can hardly conflict with the ego. Query: is ego purposive? Freud's definition of ego as space-time-cause coordination really implies a negative answer to this as well. But, clearly Freud might include in the ego the conscious commonsense purposes one has, other than the id and the super-ego, i.e. the repressed animal appetite and moral injunctions, though obviously overlapping with both of them. And so, if ego does have aims, these may be frustrated either by the id or by the super-ego, or by the external world.

Strangely, those caused by the id are neuroses, by the external world are psychoses, and by the super-ego have no name!

The symptoms of neuroses, says Freud, are ways in which the repressed id reasserts itself. The chief symptom of psychoses, he says, referring to Meynert, is 'acute hallucinatory confusion' which is utterly unrealistic. In this condition the ego does not have a true image of the world, but a false one, built on wishful thinking, very much like in dreams. We need not expand here on the similarity of this view with that of Kraepelin, nor on the paradoxes that it gives rise to.

Freud adds here to Kraepelin's picture the idea of wishful thinking dream-world. He goes further and adds that what obscures the picture is the patient's attempt at a self-cure. Here neurosis and psychosis are alike, he suggests, and we accept this as a very important point: both in psychosis and in neurosis the patient's inadequate attempt at a self-cure is part and parcel of the disease.

The etiology is always frustration by the external world, but the pathogenic effect is either repression of the id by a strong ego (neurosis) or a disruption of the ego (psychosis).[6] The super-ego, we are told, simply represents the external world, or 'the demands of reality'. This last point is very disturbing: the ego is only a coordinator, and so can include an image of the external world – a true picture, except in psychosis – and as such it can incorporate the part of the super-ego which is descriptive and coordinating; but surely, according to Freud himself an important part of the super-ego is repressed: it is repressed because it threatens, because it induces a strong sense of guilt. And so, the picture has to be handled gingerly or else it falls apart. Freud himself confesses unease, yet he insists. The super-ego as an ego-ideal which unites parts of the external world as well as parts of the id. This is not the whole story, and Freud now introduces a new illness, neither a psychosis nor a neurosis: a conflict between ego and super-ego. Melancholia, he suggests, may answer this description.

After more hesitation and repetition Freud says we need the analogue for psychosis of the repression mechanism in neurosis. (To this we hope that we have the answer; in neurosis a symbolic fixation takes place, in psychosis an abstract one, and both lead to repressions, though one emotional, one intellectual, and both are accompanied by anxiety and the like.)

Clearly, Freud's paper is unsatisfactory. We have the id, ego, and super-ego, and we now add the external world. The chief agent of the external world is frustration. Frustration occurs to the id, first and foremost, but also to the ego. Also, however, to the super-ego or the ego-ideal. This last point was taken up seriously by Alfred Adler, Karen Horney, and Erik Erikson. In spite of all partisan expressions of contempt, Adler's contribution is now accepted by orthodox Freudians simply with no mention of its author.

Now apart from the external world's frustration of id, of ego, and of super-ego, the super-ego and the ego can and do endorse parts of the external world each in its own way, and so super-ego can frustrate ego, ego frustrate id, and if super-ego frustrates id, it is only through ego and so adds no new component. In a similar vein we can say that the world frustrates the id not directly but through ego. We have, then,

The external world frustrates

(a) the super-ego – identity crisis
(b) the ego – psychosis.

And the super-ego frustrates

(c) the ego – narcissistic neurosis

And the ego frustrates

(d) the id – neurosis.

So much for Freud's first paper on the topic. His second paper is 'The Loss of Reality in Neurosis and Psychosis',[7] also of 1924, and it begins with a reference to the first.[8] He had first said that in neurosis the ego suppresses the id, but in psychosis it distorts reality. This conflicts with Freud's own observation that neurosis, too, suffers from unreality of sorts.

This piece of self-criticism is not very strong: Freud's demarcation was, *ab initio,* aetiological and pathogenic, not diagnostic; the criticism is of the demarcation as diagnostic, and so it misfires. Freud justly rejects it.

This discussion tallies well with our view that in Freud's view the two, neurosis and psychosis, are diagnostically different only in degree. It may, however, raise a few nasty problems, such as, can we have a malady originating as a psychosis and evolving only mildly, or a malady originating and manifesting itself as a neurosis, mild at first but increasingly deteriorating so as to become a true psychosis? Clearly, if the aetiology and pathogenesis should be taken seriously there seems to be no such possibility. But we can easily remedy that: suppose the ego is so harassed by the id – neurosis – that at a certain stage it gives way and distorts reality. Indeed, all neuroses come to that. Can we not see the aetiology of psychosis at times in the neurosis itself?

Freud, however, takes a different route: the mechanisms by which neurosis and psychosis evolve differ from each other. If so, then there is an essential difference between the two, as yet unspecified. Freud tries to specify: whereas the repair mechanism in neurosis is repression, even repression of (traumatic) information (e.g. amnesia), in psychosis it is procuring an adequate substitute for reality, i.e. hallucination; yet both repression and hallucination fail to solve the problem and lead to further aggravation of the problem and so on. And so, whereas a neurotic may shun a piece of reality, a psychotic invents one, and both have the trouble spread. Here Freud speaks of 'analogy': the two become separate diseases. Yet even here 'the sharp distinction' fails, since neurosis, too, feeds on a world of fantasy. But, again, there is a difference: the neurotic's fantasy only adds a hidden symbolic meaning to a piece of reality, whereas the psychotic's fantasy comes to replace a piece of reality.

This, now, brings us back to Kraepelin and to the paradox of paranoia, but otherwise we do not disagree. Rather, we wish to note two noble but conflicting tendencies in Freud: to demarcate psychosis from neurosis, as well as to equate them on the level of bare essentials.

From this we come to Freud's 'Fetishism'[9] of 1927. There, after analyzing a case of fetishism, Freud refers to the two previously mentioned papers[10] as developed 'along quite speculative lines' (p. 155), adding 'soon after this I had reason to regret that I had ventured so far.' Though he intended to give up his speculation, certain cases have brought him back to his speculation. The cases are those of 'controlled schizophrenia', to use a term coined by Arthur Koestler *(The Sleepwalkers)*:[11] the patient has two contradictory views on one specific matter, which exist side by side; and he oscillates from one to the other and back; the one view is more realistic and less desirable, the other more wish-fulfilling and less realistic. In fetishism, adds Freud, the same split can be seen, and the same obsessive neurosis; indeed, the fetish offers a reconciliation, or rather different (and conflicting) reconciliations between the conflicting views.

Freud does not explain why he had regretted his speculations concerning the demarcation between neurosis and psychosis, nor what this has to do with his analysis of some obsessions, including fetishism, as neurotic yet controlled splits. The conclusion of the paper trails off to an allusion to a criticism which Freud had launched against Adler. This can be analyzed as a projection of unease, confessed a few paragraphs earlier ('I had reason to regret'), suppressed (no explanation of the regret) and repressed and surfaced as an annoyance at Adler – perhaps because the conflict between super-ego and reality that is required in 'Neurosis and Psychosis' but is not there could be so easily made to accommodate Adler's views, had he not been such a rebel.

It may be unfair to psychoanalyze Freud's text. But what else is a commentator to do when Freud both owns and disowns his speculation? Evidently he transfers his conflict to reader and commentator alike:

if you disapprove of one move you have nastily overlooked Freud's advocacy of the opposite move, whereas if you show aversion to the conflict you do not see that the situation is delicate and you ask for too much.

So we do not mind what may be said of a critic or analyst of Freud's text. We admire Freud's conflict as we see merit both in his view of psychosis as essentially a severe neurosis and in his view of it as essentially a different ailment. Yet, logically, psychosis cannot be both essentially the same as and essentially different from neurosis. Freud has not decided the issue. But times have moved and we may try to do better than Freud. We think we have. Let us merely repeat Freud's final judgment on the topic, in his late, unfinished, posthumous, ([1938], 1940), *An Outline of Psycho-Analysis,*[12] which Ernest Jones refers to as a 'production of those last years' and judges it as 'in fact of considerable value'.[13] There, we remember (see beginning of this Appendix), Freud speaks *à propos* of fetishisms, of the essential similarity between neurosis and psychosis as seen in the fact that though fetishism is neurotic, it contains a controlled split. There is one more posthumous relevant paper, 'Splitting of the Ego in the Process of Defense'.[14] As the editor notes (SE 23, 1975; pp. 273-4), it is very interesting both in characterization and in reference to other works. He characterizes it by saying it 'extends the application of the idea of a splitting of the ego beyond the case of fetishism and of psychosis to neuroses in general'. Clearly, both Freud and the editor stuck to the last to the opinion that psychoses are neuroses of sorts. The references in the editor's note are in part to early works of Freud: the idea that psychoses are neuroses of sorts runs through the whole of Freud's career with the two noted exceptions which he 'had reason to regret'.

CHAPTER 9

CONCLUSION

TOWARDS A GENERAL DEMARCATION OF PSYCHOPATHOLOGY

We wish to close this study with a theory which demarcates[1] neurosis from psychosis. We are not speaking now of clinical diagnosis but of a general psychopathology.

Before we start, we want to define our terms, namely 'dissociations' and 'integrative principle'; all other terms are used here in full accord with common psychiatric usage. It is not easy to say exactly what professional common usage is, except to say that is it usually derivative from original technical usages introduced by the giants of the profession, particularly Kraepelin, Freud, and Piaget, but never in purist or original form. Yet we cannot possibly employ Freud's terms in their exact intended meanings, since their meanings were initially influenced to a large or a small part by the fact that Freud was an associationist whereas no one is quite an associationist today, and almost no one is anyway. But let us leave common terms and speak of the two we use repeatedly. Obviously, whatever our thinking processes are, they surely integrate to some degree, and at times they disintegrate. Symptoms and signs of thought disintegration are very well known even to common people. What the disintegration – or dissociation – itself is, we are all rather hazy about. The best model we have to date, namely Jackson's, does not pertain to thought, whether integrative or dissociative, but to the correlation of some dissociation in a high integrative center of physiological control with the integration or re-integration of a lower one.

We say nothing more on disintegration or dissociation, then, except that it is the opposite of integration. Integrative principles enter philosophy from its earliest stage; they are sometimes known as metaphysical principles, sometimes as the axioms of given metaphysical systems. Some philosophers have contended that the integrative principle of science is scientific method or even the language of science, whereas metaphysics should be ousted. Others insist that science is guided by metaphysical integrative principles which serve as their foundations.

Only recently has it been suggested that there exist varieties of integrative principles for each science and even sub-science, and that there is such a variety of integrative principles for sciences as a whole. The concept of ideology was then seen as an integrative principle, conducive to scientific research or not, as the case may be, and depending on the views at hand. Integrative principles, then, were sought and found everywhere. Thus, the famous new concept of Thomas S. Kuhn,[2] namely that of the paradigm, was also interpreted by some famous thinkers, e.g. Imre Lakatos,[3] to be metaphysical integrative principles conducive to scientific research.

All this suits us very well. If the reader gets the impression that from all this one may perhaps gain only the flavour of the integrative principle, but not what it is, then this is satisfactory also: the paranoic too may have only a whiff of an integrative principle, in lieu of one. If, however, the reader has in mind something like Cartesian metaphysics, or the three laws of Newtonian mechanics, or the Darwinian approach to biology, then this too is satisfactory.

(1) We begin with the axiom that dissociation is never primary (in Jackson's sense) but always, when it happens, secondary to, or perhaps complementary[4] to, the faulty integrative principle.[5] Consequently we reject the current view that dissociation is always psychotic. Whether dissociation takes place or not is merely a matter for the severity of the case. Whereas in neurosis any possible dissociation is a mere outcome of the severity of the suffering, dissociation in psychosis is more complex an affair, as at times it is related to some specific integrative principle which is at the root of the trouble; in psychosis, dissociation may result either from the patient's suffering as in neurosis, or from a faulty integrative principle (à la Jackson).

(2) We continue with the view of regression[6] as a version, or a part, of dissociation. It is further possible that a regression is secondary to a dissociation since, according to Jackson's principles, the patient's integrative principle, operating on a dissociated person, releases some integration on a lower level which merely looks regressed, since the lower level of integration has, in fact, been achieved by the patient in adolescence or even childhood.

(3) We further assume, with Freud, that all patients suffer from anxiety: otherwise they are not mental patients.[7]

(4) We further endorse Freud's theory of fixation more or less as it stands: fixation is a primary focus of the anxiety in particular and of the illness in general; it is integrative; it is problematic; it is emotional; it is a projection of a problematic emotion.[8]

(5) However, whereas Freud speaks only of the fixation on a concrete object or class of objects, we recognize the fixation on an abstract object (this contrary to Freud). Whereas the concrete object of fixation, according to Freud, becomes a symbol (in Freud's sense) and its emotional contents repressed or suppressed, the abstract one becomes concrete-like and hence indestructible, and hence accepted by the patient as a matter of course as valid or as true.[9] But the abstract object of fixation is not necessarily symbolic, and when it is not, it need not be suppressed. Its emotional undertone (of fear) becomes a part of the emotional color of the abstract object. Thus the abstract object of fixation may, when concretised enough, become even a symbol in Freud's sense of emotional fixation. In this case the patient may well 'forget' his integrative principle, and unlike in neurosis, will lose all compass, since, in addition, he has projected and then introjected his anxiety. When he did this, he must fear his own destruction as the end of the world and not know what to do: in brief he becomes catatonic. Moreover, the Federnian confusion of self and object is a mere corollary of projection followed by introjection or internalization, i.e. repression.[10]

(6) We distinguish only in degree[11] between diverse forms of neurosis and psychosis: there is one qualitative difference between the two, but neither comes pure in fact. That is to say, we see both emotional and intellectual impairments in both neurosis and psychosis; yet we see the fixations of the two as qualitatively quite different: concrete and abstract respectively.[12] To repeat, the impairments are matters of degree, the fixations are not. Now, it follows from this that, contrary to Freud, the neurotic's view of his own condition is impaired, i.e. below his own intellectual level. We endorse this conclusion: the neurotic refuses to see the obvious: even Freud knew that a neurotic may easily see through the same neurosis he suffers from, when he sees it in others. Yet, whereas the intellectual impairment of, say, a hysteric, is very confined, in contrast with his diffuse emotional disturbance, the opposite is the case with the psychotic, whose in-

tellectual impairment is bound to become increasingly diffuse, in contrast with his concentrated anxiety.

(7) Conceptually, the mechanism of both neurosis and psychosis is that of unsuccessful self-treatment that is both the result and the cause of the fixation. This very fixed self-treatment is a secondary mechanism of making integrative the very illness one wishes to cure oneself of – this is so in the repetition, in compulsion, in repetition-compulsion, in fixation, etc. All this is but the reintegration of an integrative principle.

(8) This explains Szasz' idea of the way patients request help. The neurotic asks some quaint request in a normal language simply because his request relates to rather idiosyncratic symbols instead of to recognizable emotion. The psychotic uses an abnormal language since his fixation is intellectual. To be more precise, his language is not in the least abnormal (here Szasz was lead astray by the irrationalism of Ludwig Wittgenstein who insisted that there can be no private language[13] in a rather dogmatic manner and tried to make it a chief philosophic integrative principle), but normally refers to abnormal ideas. These abnormal ideas are normal both within their own context (in Evans-Pritchard's sense)[14] and when translated properly.

(9) The paranoid psychotic episode[15] has to be explained psychopathologically as a borderline case: under great emotional strain, we contend, the allure of a fixation as an escape increases, and during the trial period, period of attempted fixation, the private and the public domain get mixed or confused as in our definitions of paranoia proper, yet the patient has not disintegrated or got so fixed that he cannot reconsider and so emerge out of the episode.

(10) A person may have an anxiety as well as a fixation on an idea, and a connexion of the two; he may then be classed psychotic and certainly we may wish to extend help to him; and yet he need not be a paranoic. For that he needs a fixation to which he is blinded and so view his peculiar *idée fixe* as a matter of course. Such a case, pseudo-paranoic with self-insight, has no place in traditional classification, yet he exists. He is one of the many borderline cases which our demarcation naturally suggests.[16]

(11) This point is relevant, we feel, also to borderline cases between hysteria and paranoia, including simple cases of multiple personality and even cases of regression. The regressions may be severe, as some

classical ones, or minor, such as the case of the miser. As long as the money the miser hoards is concrete and a symbol of unrequited infant love or some such, it is neurotic, à la Freud; yet when made abstract it becomes paranoic, and we need do hardly more than mention the possibility to conjure the image in the reader's mind.

(12) Depression, likewise, is a symptom of either neurosis or psychosis:[17] when a psychosis it is pessimistic, and in a peculiar way. It is manifestly psychotic when paranoic, i.e. when the pessimstic integrative principle is taken for granted and as public. (Likewise hyperactivity may be only due to a neurotic hyper-tension.)

(13) We consider any psychotic episode,[18] any recurrent psychosis, etc., in the same light: while under psychosis the patient has an abstract fixation. This, again, is more manifest in episodic paranoia, be it recurrent or not.

(14) We consider the manic depressive state a complex condition. It may be psychotic, i.e. fixed on certain ideas – pessimism or impotence as expressing depression, and over optimism or even omnipotence as expressing the manic euphoria.[19] Moreover, these may be utterly taken for granted and considered public; which will qualify the patient as paranoic. It is this very inability (during the episode) to think any other way than either his manic way or his depressive way which is the fixation which he takes for granted and will not allow criticism of. The emotional expression of mania and of depression are thus subsidiary to the thought impairment even though the diagnosis may be affective. The interesting complications are the alternation, especially when rhythmic, i.e. when phased regularly. Sometimes they seem triggered. It is not even clear that between two attacks on one pole – depression or euphoria – there comes an attack on the opposite pole – euphoria or depression (unipolar and bipolar psychoses). Yet, clearly, patients are at times self-aware, both regarding their emotional states and regarding the rhythm or trigger.[20]

(15) All cases of self-manipulation during any manic-depressive cycle must be classed as neurotic. Such patients, unlike those suffering from manic-depressive psychosis proper, can work their way up in order to reach down, or the other way round, often using imagery like roller-coaster, in a semiconscious – hysterical – manner. All this does not hold in the least, of course, for cases when the awareness of one's

emotional condition is entirely excluded during the manic-depressive episode, which marks the episode as psychotic par excellence. Finally, the condition may become dissociative, and on either pole. But, as we said, dissociation is merely an outcome. In this case, strangely, it is the very integrative principle on either pole, run to the extreme, which itself dissociates the patient: in mania into word salads and such, in depression to sheer stupor. The very powerful logic of this condition is, again, paranoic par excellence. This makes the intellectual component of this affective disorder quite prominent and easy to observe. The reason it was thus far overlooked, is, of course, that the emotional component is very prominent, contrary to the classical views of *paranoia vera.* (One who did not overlook it, however, was, of course, Binswanger.[21])

(16) Already Bleuler presented schizophrenia as primarily an intellectual impairment;[22] he considered it looseness of association, but we need not consider associationism at all and simply update his picture by replacing the looseness of association with a looseness of the thinking process without any further specification of what this process is in general. Bleuler's idea is hardly heterodox these days. Federn's view of the psychotic's fusion of his body with the world or Laing's variant of it as the schizophrenic's alienation from his body we consider not necessary but one possible expression of such looseness – perhaps under conditions described above. Laing's profound observation that the same view may be expressed by a sane Cartesian philosopher and by a schizophrenic is to us an example of paranoia in Laing's schizophrenic patient, who, unlike Descartes, takes the Cartesian view to be the same as the common sense normal one (whereas, as Freud has noticed in his *The Ego and the Id,* commonsense says both that I dwell in my body and that I am my body).

(17) We consider many diagnoses of schizophrenia to be misdiagnoses of paranoia in the psychopathological sense, and regard the dissociative signs, which in acceptable practice preclude the diagnosis of paranoia, as mere advanced signs of the illness which should not preclude this diagnosis. Whether we call it paranoia or schizophrenia matters little.

(18) The original source of any psychopathological condition may be a trauma[23] or a chemically induced fear or any other event or state

of affairs or series of these or even nothing at all. What we insist on is that the illness is not the source of the fixation or even the fixation itself but merely – to echo Freud again – the effort to adjust to life which as a by-product makes life even harder: it intensifies the fixation and makes it both central and integrative. This is the application of Freud's insight to psychotic cases. He says explicitly that whereas most adolescents shake off certain fixations, others intensify them and so become psychopathological. This amounts to the admission, all his statements to the contrary throughout his career notwithstanding, that psychopathology as he envisaged it, is adolescent and post-adolescent.[24] Freud's own claim that psychopathology is infantile must therefore be understood differently now: the seed of psychopathology is, according to Freud, infantile; yet its onset must be post-adolescent. Similarly, we must reread Klein's view. In the sense that Freud's infant is neurotic, Klein's is psychotic: yet both agree that in adolescence this may be remedied or else become psychopathological. And so, we do not comment on Freud's or Klein's theory of the infantile source of the illness; we only declare all mental illness, neurotic or psychotic, to be post-adolescent. This is not to deny that both infants and adolescents have emotional problems akin to neurosis or to psychosis and at times are in need of help. Yet, at least if Piaget is not to be dismissed off-hand, these must be declared of a different order.[25] For, Piaget relates the child's emotional experiences to his level of intellectual ability, and these are variable and must be taken into account when studying infant and adolescent emotional problems.

(19) We consider cure from a psychopathological state, whether temporary or permanent, or even the seeming transition from one disease to another, only changes in states: we have not outlined any dynamic theory,[26] or any mechanism, of going into or out of a psychopathological state. We merely demarcate states.[27] In particular, we also allow for, and demarcate, mixed states.[28] That is to say, we allow for cases where neurotic symptoms occur in psychotic patients and vice versa.[29]

(20) To conclude, we do not know what illness is, physical or mental; what is health;[30] we do not offer any etiology,[31] and we do not offer any prescription for treatment; all we offer here is an extension of Freud's theory of neurosis as a fixation which is both illness and

self-treatment; we join Ey in considering both neurosis and psychosis, both integrative and dissociative.[32] This essay centered on the structural defect of the intellectual apparatus, as expressed in paranoia, and attempted an extension of its conclusions into a theory of demarcation of psychopathology.

POSTSCRIPT

In our opinion the following is what signifies most in the matter. What is common to both neurotics and psychotics is not only a fixation but also a blind-spot – which isn't really blind: under immense pressure the patient may show recognition of what he systematically ignores. Indeed ignoring it costs so much intellectual and emotional energy and recognizing it costs enormous emotional energy – so much so that it is a threat of total collapse.

Hence, all fixation is a form of mental block, where a block is, as already noted by Anna Freud and more so by Melanie Klein, an active defence principle. Moreover, in our opinion it is so active as it is no less preventive of one integration as supportive of another – this in full justice to Sigmund Freud's most important insight of mental illness as a defective yet self-reinforcing self-treatment. We only add that the self-treatment has moments of self-assertion, of self-expression, of self-realization, even a strong sense of identity and a sense of freedom – however misplaced and pathetic these may be. This is in justice to Ey who stresses that all mental illness is an illness of freedom.

Now, the extreme case is the paranoic. He has a fixation on one abstract system and a block on the common system: he pretends to be blind to it. Hence, in a simple sense he suffers from a split-personality, as described in the vernacular and folk-mythology. Hence, paranoia and schizophrenia as analysed today clinically are originally of a common origin with the onset of dissociation up to dementia – which we consider a mere matter of the course of the illness. It is rather a mistake that dissociation is taken to be more characteristic of schizophrenia rather than of paranoia; it is only more powerful.

We leave it for further study to consider diverse combinations of neurosis and psychosis, as well as of the diverse neuroses and psychoneuroses.

NOTES

CHAPTER 1

[1] The etymology of the Greek word 'paranoia' is para = besides; nous = thought, mind. It is the same as that of the French word, 'délire', which, from the Latin is de = besides; lira = furrow. 'Schizophrenia', etymology in Greek means, schizo = split; phren = mind. The origin of the word phren for mind, is from the appelation of the nerve (phrenicus) which regulates diaphragmatic movements. The Greek saying was: "Respiration is the mirror of the soul," very much like the Hebrew etymology for the soul, which is 'nefesh', meaning breath, 'neshama', meaning the same, 'ruah', meaning wind, and the etymology of the Latin 'spirit'.

[2] The extremely low frequency of paranoia is utterly insignificant for us. Indeed, were *paranoia vera* non-existent, we would invent it by idealization. We would consider in abstract the paranoid syndrome, so common in schizophrenia and present in other cases too. What concerns us is the problem.

[3] The fact that Kraepelin saw only 19 paranoics throughout his whole career is recorded fifty years later by Mayer-Gross. See W. Mayer-Gross, 'Psychopathology of Delusions: History, Classification and Present State of the Problem from the Clinical Point of View', pp. 59-87, in F. Morel (ed.), *Psychopathologie des délires,* Hermann, Paris, 1950: "Kraepelin based this rare entity on a small sample, 19 patients in all. When Kolle [Kraepelin's disciple] went out in 1931 to collect similar cases in all German mental hospitals (in order to include these with his own), he found only 47 satisfying the definition" (p. 80).

[4] Since Freud's famous analysis of the Schreber case and his suggestion that paranoia is a result of repressed latent homosexuality, many other serious emotional troubles have been suggested as a possible source of the disease, thus viewing paranoia as a cover for all sorts of serious emotional troubles. To begin with, see D. P. Schreber, 1903, *Memories of My Nervous Illness* (transl. by I. MacAlpine and R. Hunter), Dawson & Son, London, 1955. Then, see S. Freud, 1911, *Psychoanalytic Notes on an Autobiographical Account of a Case of Paranoia (Dementia Paranoides),* Hogarth Press, London, *S.E.* 12 (1975), 3-82; S. Freud, 1915, *A Case of Paranoia Running Counter to the Psychoanalytic Theory,* Hogarth Press, London *S.E.* 14 (1975), 261-274; and S. Freud, 1922, *Some Neurotic Mechanisms in Jealousy, Paranoia and Homosexuality,* Hogarth Press, London; *S.E.* 18 (1975), 221-232.

By contrast, Rycroft sees in latent homosexuality not the cause of paranoic anxiety but a defense mechanism against it. See C. Rycroft, 'Two Notes on Idealization Processes', *Intl. J. Psycho-anal.,* 36 (1955), 81-87, reprt. in his *Imagination and Reality,* Hogarth Press, London, 1968, pp. 29-41.

For various problems other than latent homosexuality in paranoia, see S. Ferenczi (1914), 'Some Clinical Observations on Paranoia and Paraphrenia', in his *First Contributions to Psychoanalysis,* Hogarth Press, London, 1952; V. Tausk (1919), 'On the Origin of the "Influencing Machine" in Schizophrenia',

Psych. Quart. 2 (1933) 519-556, reprinted in R. Fliess (ed.) *The Psychoanalytic Reader,* Hogarth Press, London, 1948, vol. 1, pp. 31-64; J. H. W. von Ophuijsen, 'On the Origin of the Feeling of Persecution', *Intl. J. Psycho-anal.* 1 (1920), 235-239; R. P. Knight, 'Relationship of Latent Homosexuality to the Mechanism of Paranoid Delusions', *Bull. Menninger Clinic* 4 (1940), 149; K. Menninger, *Love Against Hate,* Harcourt, Brace, New York 1942; J. A. Arlow, 'Anal Sensation and Feelings of Persecution', *Psycho-anal. Quart.* 18 (1949), 79-84; R. C. Bak, 'Masochism in Paranoia', *Psycho-anal. Quart.* 15 (1946), 285-301; I. MacAlpine and R. Hunter (1955), their discussion chapter, pp. 369-411, of Schreber's *Memories* (1903); W. R. D. Fairbairn, 'Considerations Arising Out of the Schreber Case', *Brit. J. Med. Psychol.* 29 (1956), 113-127; C. Rycroft, 'The Analysis of a Paranoid Personality', *Intl. J. Psycho-anal.* 41 (1960), 59; E. Lemert, 'Paranoia and the Dynamics of the Exclusion', *Sociometry* 25 (19062), 2-20; reprinted in T. J. Scheff (ed.), *Mental Illness and Social Process,* Harper and Row, New York, 1967; J. Nydes, 'Schreber Parricide and Paranoic Masochism', *Intl. J. Psycho-anal.* 44 (1963), 208-212; H. M. Wolowitz, 'Attraction and Aversion to Power: A Psychoanalytic Conflict Theory of Homosexuality in Male Paranoids', *J. Abnorm. Psychol.* 70 (1965), 360-370; J. Lacan, 'Du traitement possible de la psychose'; in his *Ecrits,* Ed. du Seuil, Paris, 1966, pp. 531-583; A. Crowcroft, *The Psychotic: Understanding Madness,* Penguin, Harmondsworth, 1967.

New views about the Schreber family are brought up by F. Baumeyer, W. G. Niederland, and M. Schatzman. For heredity in the Schreber family, see F. Baumeyer, 'The Schreber Case', *Intl. J. Psycho-anal.* 37 (1956), 61-74. About who really was Schreber's father, see W. G. Niederland, 'Schreber: Father and Son', *Psycho-anal. Quart.* 28 (1950), 151-169; W. G. Niederland, 'The "Miracled-Up" World of Schreber's Childhood', *Psycho-anal. Stud. Child.* 14 (1959), 383-413; W. G. Niederland, 'Schreber's Father', *J. Am. Psychoanal. Assoc.* 8 (1960) 492-499; W. G. Niederland, 'Further Data and Memorabilia Pertaining to the Schreber Case', *Intl. J. Psycho-anal.* 44 (1963), 201-207.

For a complete reversal of Niederland's views see M. Schatzman, 'Paranoia and Persecution: The Case of Schreber', *Intl. J. Psychiatry,* 10 (1972), 53-78; and M. Schatzman, *Soul Murder: Persecution in the Family,* Allen Lane, London, 1973.

Another dimension altogether, a moral one, is introduced by H. Baruk and A. Hesnard. See H. Baruk, 1945, *Psychiatrie morale, expérimentale, individuelle et sociale. Haines et réactions de culpabilité,* Presses Universitaires de France, Paris, 2nd ed., 1950; A. Hesnard, *L'Universe morbide de la faute,* Presses Universitaires de France, Paris, 1949.

Finally, see an abstract of a symposium on 'Reinterpretation of the Schreber Case', *Intl. J. Psycho-anal.* 44 (1963), 90-223.

In spite of all these works, the tendency of most psychoanalysts is to consider Freud's insight into paranoia as valid.

[5] There are surprisingly few studies devoted solely to paranoia in the psychiatric literature, and most of them are by French authors. Let us also point out the fact that these few are spread over a period of almost 100 years. Those which are available, however, are of outstanding quality. See J. Séglas, 'La paranoïa', *Arch. Neurol. Psychiat.* 13 (1887), 62-76, 221-232, 393-406; P. Sérieux and J. Capgras, *Les folies raisonnantes: le délire d'interprétation,* Alcan, Paris, 1909; A. Binet and T. Simon, 'La folie systématisée', *Année Psychologique,* 1910;

E. Bleuler, *Affectivity, Suggestibility, Paranoia,* State Hospital Press, Utica, New York, 1912; See also *State Hosp. Bull.* 4 (1912), 481-601; J. Génil-Perrin, *Les paranoïaques,* Maloine, Paris, 1926; J. Lacan, *De la psychose paranoique dans ses rapports avec la personalité,* le François, Paris, 1932; F. Morel (ed.), *Psychopathologie des délires,* vol. 1 of the *Proceedings of the International Congress of Psychiatry,* Herman, Paris, 1950. It includes works by P. Guiraud, 'Pathogénie-étiologie des délires'; W. Mayer-Gross, 'Psychopathology of Delusions: History, Classification and Present State of the Problem from the Clinical Point of View'; E. Morselli, 'Recherches expérimentales et délires'; and H. C. Rümke, 'Significance of Phenomenology for the Clinical Study of Sufferers of Delusions'; H. Ey (1953), *Les délires (généralités)* cours (révu 1967), mimeograph. D. Shapiro, 'The Paranoid Style', pp. 54-107, in his *Neurotic Styles,* Austen Riggs Center Monograph Series, No. 5 Basic Books, New York, 1965.

For remarkable works, though not directly concerned with paranoia, but very pertinent to the problem of delusions, see E. Kretschmer, 1918, *Der Sensitive Beziehungswahn,* Springer, Berlin, 2nd ed., 1927 (transl. *Paranoïa et sensibilité,* Presses Universitaires de France, Paris, 1963); W. Mayer-Gross, *Selbstschilderungen der Verwirrktheit. Die Oneiroide Erlebnisform,* Springer, Berlin, 1924 (parts of which are translated in H. Ey (1948-1954), *Etudes psychiatriques,* Desclée de Brouwer, Paris, vol. III, 1954; pp. 250-279); S. Follin, *Les états oniroïdes,* Masson, Paris, 1963.

[6] See Hippocrates, *The Sacred Disease* (transl. W. H. S. Jones), Loeb Classical Library, London and New York, Vol II, 1928; pp. 139-141. "The disease called 'sacred' is not, in my opinion, any more divine or more sacred than other diseases, but has a natural cause and its supposed divine origin is due to men's inexperience and to their wonder at its peculiar character".

The word for sacred in the original means not necessarily the holy, but also possibly the unholy – as all words for taboo (including 'taboo') are.

Perhaps Hippocrates' repudiation of the view of epilepsy (i.e., mental illness) as divine is his best known line. Perhaps we pay him this homage in order to conceal the embarassing fact that we refused to pay any attention to his repudiation until the French Revolution. Ordinary histories of psychology or of medicine sound odd in their praise for both Hippocrates and his disciples of the Enlightenment, since they do not explain what Pinel has added to what Hippocrates said. He added, of course, an attempt to apply Hippocrates' view.

[7] Mental illness in modern times is no longer ascribed to supernatural causes; it is considered as an accepted disease. Is this really so? The answer is no – even today mental illness is looked upon as a disease, yet in a category of its very own. One still has to fight in order to really liberate the mentally ill, to release them from their confinement and their bondage, be it a sociological or a pharmacological straightjacket. The anti-psychiatry movement of our day is an extension of the antiasylum tendency which began after World War II.

[8] T. Szasz, in *Law, Liberty and Psychiatry,* MacMillan, New York, 1963; p. 2, says: "On May 24, 1794, Philippe Pinel removed the chains from one of the most feared patients at the Bicêtre, the Paris asylum for male lunatics. Thus the historical paradigm of psychiatric treatment is neither prescribing medicines nor performing operations, but giving an imprisoned human being a measure of freedom. Esquirol and Ferrus, the outstanding French psychiatrists of the first

half of the nineteenth century, were both students of Pinel. More than anything else, they were 'prison-reformers'." This panegyric is heartwarming, yet in fairness we may remember Pinel, 1801, *Traité médico-philosophique de l'aliénation mentale, ou la manie,* Brosson, Paris, 1809; (transl. *A Treatise on Insanity,* Hafner, New York, 1962) as well as E. Esquirol's *Des maladies mentales,* Tircher, Bruxelles, 1838; (transl. *Mental Maladies,* Philadelphia, 1845); both of which offer therapeutic measures of different kinds, including moral treatment for the mentally ill. It is one thing to see Pinel and Esquirol in the light of their given society, and to criticize them (together with their society's views) and altogether another thing to see them as predecessors to Szasz and the humane movement.

A similar view to Szasz' is expressed by M. Foucault. In his 1961 *Histoire de la folie à l'âge classique,* Gallimard, Paris, new ed. 1972 and in several places, both implicitly and explicitly characterized Pinel not only as a speudo-liberator, but indeed as a guardian of society. See especially Chapter IV of Part 3. (The transl. of 1961 ed., *Folie et déraison,* Plon, Paris, as *Madness and Civilization: A History of Insanity in the Age of Reason,* Pantheon, New York, 1965.)

9 The greatness of P. Federn should not be overlooked regardless of the terrible confusion of the Federnian literature. First, Federn was the first to demarcate psychosis as totally of a different order from neuroses – and this despite his close association with Freud. Second, he had a theory which is refutable (and refuted) about the confusion of self and object as the cause of psychosis. Third, he described states in which normal people may feel the same confusion. He thus was the first to describe something like psychosis in everyday life. See P. Federn, *Ego Psychology and the Psychosis,* Basic Books, New York, 1952.

10 "You will realize further significance of the insight so acquired when you learn that the mechanism of the dreamwork is a kind of model for the formation of neurotic symptoms". S. Freud, 1916-1917, *Introductory Lectures on Psychoanalysis,* Hogarth Press, London, *S.E.* 15 and 16 (1975); vol. 15, 11th Lecture, p. 183. Compare with J. E. D. Esquirol, who in 1838, humanizes the madman yet at the expense of making madness ever so more repulsive: his monumental *Des maladies mentales,* begins thus: "What meditations for the philosopher who, wanting to avoid the tumult of the world, passes through a home of madmen! He will find there the same ideas, the same errors, the same passions, the same misfortunes: it is the same world: but in such a home (of madmen) the traits are stronger, the nuances more marked, the colours are livelier, the effects more striking, because man is there in all his nudity, because he does not dissimulate his thought, because he does not paint his passions with the charm of seductiveness, nor his vices by the appearances which mislead". E. Esquirol, *Des maladies mentales,* Tircher, Bruxelles, 1838; vol I, p. 1.

11 That delusion can be systematic is ancient knowledge and central to Cervantes' popular *Quixote.* This type of delusion was presented by Kant in his *Anthropologie,* cited extensively in note 15 to Chapter 5 below. It was modified by diverse writers from Esquirol to Bleuler. What is so astonishing is that Kant has it all – the delusion obsessively sustained with great logical acumen coupled with a complete disregard to public opinion *(sensu communis).* Yet his followers, attempting a closer accord with the facts got distracted. First came Esquirol, who distinguished three main types of *monomanies: intellectuelle, affective,* and

instinctive. His *monomanie intellectuelle* is the closest to the present day definition of paranoia. "The patients start from a false principle, which they follow logically without deviation, and from which they derive legitimate consequences which [in their turn] modify their affects and acts of their volition; outside this partial delusion, they feel, reason and act as everybody else." (E. Esquirol, *Des maladies mentales,* 2 vols., Tircher, Bruxelles, 1838; vol. 1 p. 332, our translation.) This classification raises the question, is paranoia primarily intellectual or primarily affective? It was decided differently in the different editions of W. Griesinger's leading *Die Pathologie und Therapie der Psychischen Krankheiten, für Artzte und Studierende,* of 1845 and 1861: only the latter decides in favor of the intellectual defect theory. But before this could at all be properly examined, paranoia had to be classed apart from all other mental disturbances and illnesses, as was indeed done by Kraepelin. That is to say, Kant might be seen as so astonishingly precise about paranoia, as we would think today, only because he overlooked everything else. The process of singling paranoia out was lengthy: one should realize that unlike Kant, others tried to cope with it, and they did so by classifying it together with other and similar illnesses. Kraepelin kept it apart, and concurrently J. Séglas and his followers. Yet by and large it is Bleuler's view which is widely accepted, and it presents paranoia as a form of schizophrenia. Needless to say, for our own part, as we take paranoia to be the limiting case of all psychosis, we cannot take part in this dispute. But clearly, it had a point, and so Kant's description was rightly deemed not very helpful.

[12] The word 'paradox' usually means in English a seeming contradiction; for example in *The Pirates of Penzance,* the young man paradoxically had only six birthdays though he was twenty-four. More specifically a paradox looks at one and the same time both obviously true and obviously false, as many witticisms of sharp tongued aphorists are, from LaRochefoucauld to George Bernard Shaw.

In logical texts often the word 'Paradox' is used as synonymous with 'antinomy' to mean proven contradictions. Let us note here, by the way, that paranoics may have slight alternations of meanings of words systematically employed, to render a view contradictory at times (so as to enable them to deduce whatever conclusions they wish) but only seemingly contradictory at other times (when their views are under scrutiny); there is nowhere anything like a logical paradox in our discussion. (See note 8 to Chapter 8 below.)

[13] Some direct attacks on the concept of 'the logically reasoned delusion' from the point of view of the paradoxicality (or of the 'logical scandal') which this concept presents, have been undertaken by K. Schneider (1931), *Psychopathologie der Gefühle und Triebe im Grundiss,* G. Thieme, Leipzig, 1935, quoted by H. Ey, 1948-1954, *Etudes psychiatriques,* Desclée de Brouwer, Paris, vol. 2, 1954, p. 525, note. In this context Ey refers also to C. Schneider and to O. Kant. It is H. Ey, who in his course *Les délires (généralités)* 1953 (rev. 1967) puts most of the paradoxes together.

The clearest concise statement we could find of the paradoxes, and even a near solution to them is, we think, Rycroft, C., 1968, *A Critical Dictionary of Psychoanalysis,* Penguin, Harmondsworth, 1972.

"Delusion. Term used in psychiatry to refer to a belief, voiced by a patient, which is both untrue and uninfluenceable by logic or evidence; a fixed idea. A common-sense, clinical concept which turns out to be unexpectedly difficult the moment one asks two questions: (a) How does the psychiatrist know his corresponding belief is true? (b) In what sense does the patient believe his delusions? (a) can be answered only if it is possible to discover the function of the beliefs in the mental economy of the patient and the psychiatrist. (b) leads to the conclusion that the delusion is a manifestation of a thought-disturbance, viz. loss of the capacity to distinguish between categories of thought – in the simplest instance, between metaphorical and factual statements. If the correct mode can be discovered, delusional ideas often can be shown to make sense." pp. 28-29.

No doubt the diagnosis of some cases of paranoia as a confusion between a metaphor (or simile) and a fact, plays a central role; but in others it is a concretization and in others a simpler error. (See note 22 to Chapter 9 below.)

[14] Our remarks on paranoia are acceptable even to those who do not consider paranoia as a separate entity, but as one of the symptoms of schizophrenia, though, of course, they would have to be put in their own terms. We speak specifically of the case where there is a "systematic delusion in which clarity, volition and order of thinking are preserved intact." Now most physicians report having seen such cases here or there. They usually disagree, however, about two things. First, most of them see it as a symptom, not as a primary defect. Second, they view it differently as to where exactly it takes place in the course of the disease. What we wish to stress here, is that the clinical observation is not in dispute, nor is the definition of paranoia, only the diagnosis. Hence, the paradoxes arise regardless of our classification, and therefore we have no objection to classifying them in this or that way and have this reflected in their names: for example, one may call them the paradoxes of schizophrenia.

[15] When we speak of paranoia as an impairment of the thought-process we immediately raise the question of demarcating it from other such impairments, whether caused by apahasia, alcoholism, or other means. We do not mean to class all thought-process impairment as paranoia, and we shall later explain which impairment is. Yet we can already say, whereas everybody from Kant to Kraepelin and Freud and Bleuler stressed the delusion and denied impairment of thought-process, indeed they all noted the contrary, namely the logical power of the paranoic and his ingenuity, it is this very quality that we take as the primary sign of an impairment of the thought-process itself.

[16] We did not yet mention Hughlings Jackson's principles. See further in the text. Suffice it to say now, that H. Jackson recognizes a local defect, as a defect nonetheless. Speaking of convulsions, he says there is a difference between a general and a local convulsion, "not merely in degree of more or less spasm, – but also in degree of evolution of the nervous processes which are unstable", and "I trust I am studying the general subject of convulsion methodically when I work at the simplest varieties of occasional spasm I can find." ('A Study of Convulsions', 1870, in *Selected Writings,* (ed. by J. Taylor.), Hodder and Stoughton, London, 1931-1932; vol. I, p. 8, text and note.)

It is clear that for Jackson a local defect is, in principle, not different from a global one: both have to do with degrees of dissolution of the nervous system.

We suppose that any fixation, emotional or intellectual, renders the individual less intelligent (in Piaget's sense) than he should be; and that this is so by some Jacksonian structural principle. See note 19 below, and see J. Piaget, 1947, *The Psychology of Intelligence,* Routledge and Kegan Paul, London, 1967: "Only intelligence, capable of all its detours and all its reversals by action and thought, tends towards an all-embracing equilibrium by aiming at the assimilation of the whole of reality and the accommodation to it of action, which it thereby frees from its dependence on the initial here and now" (p. 9).

[17] In his *On the History of the Psychoanalytic Movement* (1914), Hogarth Press, London, *S.E.* 14 (1975), 7-66, repr. 1962; ch. 2, p. 416, Freud quotes E. Bleuler, *Die Kritiken der Schizophrenie,* 1914, who says, "the psychology of the depths offers something towards a psychology which still awaits creation and which physicians are in need of, in order to understand their patients and to cure them rationally . . . I even believe that in my *Schizophrenie* I have taken a very short step towards the required understanding." Freud is "content with this admission."

[18] "Thomas Sydenham, the greatest clinician of the 17th century and one of the greatest physicians of all times, distinguished two groups of symptoms: *symptomata essentialia* [primary], the symptoms caused by the lesion, and *symptomata accidentalia* [secondary], the symptoms caused by the reaction to the lesion. When a finger is burnt, tissular parts are destroyed by the action of the heat. We observe phenomena directly derived from the lesion. The atrophied tissulary parts play the part of foreign bodies in the organism which reacts, eliminates them and replaces them by new cells. Thus we observe a whole syndrome which is the expression of this reaction. According to Sydenham's conception, the clinical picture of the burn is thus composed of some symptoms caused by the lesion, and others by the reaction. In numerous cases, it is extremely difficult to say into which group one has to classify this or the other symptom". H. Sigerist, *Introduction à la médecine,* pp. 109-110, quoted in R. Dalbiez, *La méthode psychanalytique et la doctrine Freudienne,* Desclée de Brouwer, Paris, 2 vols. (2nd ed.) 1949; vol. II, pp. 278-279 (our translation).

[19] In the pathology of the process of thinking we call primary those defects of a higher level of integrative organization, which by their absence release an integrative principle of a lower level – which we call secondary. This is in an attempt at a generalization of Jackson's view.

The reverse exists in Freudian terminology. Freud considers a primary thinking process precisely the one which we view as secondary, and reversely – Freud sees as a secondary thinking process, what we would name a higher integrative principle which, when defective, becomes a primary element of the disease. See note 20 below.

Now, there is more to all this than a mere terminological difference. The point of significant difference of opinion is this. For Freud, the source of mental pathology is in its point of origin, and its origin is usually a lower level point, not a higher level point. The higher integrative principle which is destroyed and consequently 'releases' a lower level, is unnoticed by Freud. On the contrary, Freud views the source of mental pathology as always being in the content and thought-structure of the lower level, which in his opinion always actively invades and destroys that of the higher level. He named the destructions of the lower level, primary: what is primary, he said, is the universal law. The end result is the same: the destruction of the high level principles of integration and consequently free play of the low level ones. The question is only this: what took place first?

Now, for a Jacksonian the answer is simple: first comes the destruction of the higher level; for a Freudian it occurs in reverse: first comes the 'invasion' of the lower level. Which is the true picture? And why not ask: is not a synthesis of these two views possible?

Not that though we have no synthesis to offer, our view paves the way for a synthesis, since while using Jackson's principle structurally we offer no etiology and no pathogenesis. In particular, we do not exclude either a Freudian or a Jacksonian theory of pathogenesis – this in accord with a point first made and emphasized by H. Ey. Notice, incidentally, that the order or hierarchy of organizing principles is, of course, a partial order. See J. Hughlings Jackson, 'Evolution and Dissolution of the Nervous System', (Croonian Lectures), 1884, pp. 45-75 and 'The Factors of Insanity', 1894, pp. 411-421, in the *Selected Writings of John Hughlings Jackson* (ed. by J. Taylor, Hodder and Stoughton, London, 1932; compare with S. Freud, 1916-1917, *Introductory Lectures on Psychoanalysis,* Hogarth Press, London, *S.E.* 15 and 16 (1975).

[20] S. Freud, 1940, *An Outline of Psychoanalysis,* Hogarth Press, London, pp. 141-207 of *S.E.* 23 (1975); ch. 4, p. 164: "But behind all of these . . . there lies one new fact the discovery we owe to psychoanalytic research . . . processes in the unconscious or in the Id obey different laws from those of the preconscious ego. We name these laws in their totality the *primary process,* in contrast to the *secondary process* which regulates events in the preconscious or ego." (Italics in the original text.)

See note 19, above.

[21] The place occupied by Jackson's principles in the medical (and today even in the biological) literature, is widely known. It is interesting to note, therefore, that when Hughlings Jackson announced these principles they were hardly noticed. See J. Delay, *Etudes de psychologie médicale,* Presses Universitaires de France, Paris, 1953. And yet already then Jackson had a reputation as an eminent neurologist. He had described for the first time diverse neuropathological conditions (symptomatic epilepsy still carries his name). What is relevant to our discussion here, since we are at present concerned with the growth of ideas, is the following fact. Sigmund Freud was one of those who knew and appreciated Jackson's principles shortly after their publication – and even in depth. He widely used Jackson's views as stated in his earliest monograph, S. Freud, 1891, *On Aphasia: A Critical Study,* International Universities Press, New York, 1953, p. 87. Yet, in his other works there is no mention thereof, not even of Jackson's name.

[22] T. Ribot, the famous French professor of psychology in Paris was a contemporary of Jackson, and shared his views. He announced his famous 'law of dissolution' in 1876. See T. Ribot, *La logique des sentiments,* Alcan, Paris, 1905. Today the psychiatrist continuing in the Jackson tradition (neo-Jacksonian) is H. Ey. See his *Traité des hallucinations,* Masson, Paris, 1973.

[23] Even in mental deficiency (retardation) there are integrative principles in that the lower levels of intellectual organization of the patient remain intact. See B. Inhelder, 1943, *Le diagnostic du raisonnement chez les débiles mentaux,* Delachaux et Niestlé, Neuchatel, 2nd ed., 1963. See also M. Woodward, 'Piaget's Theory', in J. Howells (ed.), *Modern Perspectives in Child Psychiatry,* Oliver & Boyd, Edinburgh and London, 1965.

[24] See Chapter 2, note 22.

Incidentally, Jackson's principle in neurology is a part of a tradition within the history of medical thought. Generally, "two opposite philosophies competed for medical recognition. Some, inclined towards materialism, do not see the disease but as the passive outcome of eternal causes. For them only the negative symptoms are of importance. Their adversaries, of the vitalistic orientation, make of the disease an active principle and pay attention mainly to the positive symptoms. In psychiatry, the materialists are called 'organicists', and the vitalists, 'psychogenetists'. One wonders whether there is no place for a third view". R. Dalbiez *loc. cit.* in note 18 above.

[25] From Jackson's principles it follows, first, that damage to a higher integrative principle gives autonomy to the highest of those subservient to it, and second, that the higher principle is newer than the lower – both ontogenetically and phylogenetically. Thus, originally, Jackson relies on the famous and seductive view, now rejected, that ontogenesis recapitulates phylogenesis – though not embryologically but physiologically. Now, though in embryology the hypothesis is definitely 'out', whether in physiology it is still 'in' or not is hard to say: it is certainly even more seductive there.

[26] P. Seligman, reviewing E. Straus, M. Natanson, and H. Ey, *Psychiatry and Philosophy,* Springer (New York, 1969) writes in *Philosophy of Science* 39 (1972), 99-101, as follows: "...Henry Ey... professes indebtedness to the English neurologist Hughlings Jackson (*Elements of Madness,* 1884). Ey conceives of the normal mind as a psychic superstructure developed by an evolutionary passage from an organic infrastructure. Mental illness is then defined as a dissolution or regression, i.e. as a disorganization of higher processes and a psychic reorganization at a lower level." (p. 100).

[27] As tradition has it, B. Morel, a pupil of Esquirol, was the first one to have used the name *Dementia Praecox* in psychiatry. See his *Etudes cliniques,* 2 vols., Baillière, Paris, 1852; vol. 1, pp. 37-38, vol. 11, pp. 257-303. It is said that the first part of the term, the 'Dementia', implies both that the condition has to do with a diminution of the patient's mental state, as well as the fact that this diminution occurred to somebody who previously developed normally and has been mentally sane. Dementia may normally occur in old age; it is then called senile dementia. If it occurs at an early age, it is a 'praecox dementia'. Originally, for Morel himself, the term 'praecox' did not mean 'early' but 'rapidly deteriorating'. Moreover, it was a descriptive term, not at all one of principle. Moreover, in 1852 he used the term as a non-technical adjective, as one amongst others which he used to qualify the technical term 'dementia', during his running descriptions of diverse patients' states. His *Traité des maladies mentales,* Masson, Paris, 1860, a later work, shows even less awareness of the peculiarity of the condition later labelled hebephrenia or hebephreno-catatonic, and which he labelled "praecox". The mistaken traditional attribution to Morel of the original term "Dementia Praecox" is discussed in F. Constant-Trocme, 'Introduction à la vie et à l'oeuvre de B. A. Morel', *Confrontations psychiatriques* 11 (1973), 31-50, esp. pp. 45-47.

E. Kraepelin grouped under 'dementia praecox' the conditions previously regarded as separate entities – Catatonia (i.e., complete 'spastic' withdrawal), hebephrenia, and dementia paranoides. Kraepelin further claimed (1) that these 3 clinical conditions are interchangeable and (2) that all of them tend towards a final demential state (a complete 'flaccid' withdrawal). (As will transpire later, we think condition (2) holds for all psychoses and, given enough time, even all neuroses).

E. Bleuler viewed dementia praecox differently. He claimed that what unites catatonia, hebephrenia, and dementia paranoides (Bleuler added a fourth form: the simplex), is not their interchangeability and their demential final outcome, but the fact that all of them are characterized by the same fundamental signs: in all of them there is an impairment in the thinking process (looseness of associations), in the affect (impoverishment in the realm of feelings and in their expression), ambivalence and autism. These signs are found only in these conditions, says Bleuler, and hence his coining them as fundamental signs. All other symptoms and signs are accessory ones: They may as well appear in other mental diseases. Because of the 'split' between thinking, affect and action, Bleuler called these conditions schizophrenias.

Now, it is clear that schizophrenia, thus defined, is a mental state where a general deterioration of the patient has taken place. In psychopathological terms we speak of dissociation. Schizophrenias, then, are dissociative mental states.

Now, compare schizophrenia proper with another mental condition – paranoia. Paranoia, by definition, is the opposite of dissociation. Not only are there no disturbances in thinking, willing and deed (Kraepelin, 1893 edition), but it is also, as we have argued, a state where an integrative principle eminently plays a part.

Is this to say, then, that paranoia is not a psychosis? At least Bleuler and his (numerous) followers would have to say, no. Yet, clearly, the answer is, yes; and nevertheless we wish to retain the idea of Bleuler of psychosis as a dissociative state or condition. This is but another way of presenting the paradoxes of paranoia. Paranoia looks the very opposite of the disintegration that it is. To conclude more precisely, since Kraepelin insists that paranoia is integrative and Bleuler that it is dissociative – not in these words of course – we, who agree with them both, have a paradox at hand.

See notes 12 and 13 above and note 23 to Chapter 2 below.

[28] Some well-known psychiatrists deny paranoia the status of a separate clinical entity, and see it as part of the schizophrenic condition or state – obviously not as that clinical picture where the delusion appears immediately together with the fundamental signs, but as a special form of schizophrenia, in which a systematic delusion, logically sustained, appears together with clarity of thinking and volition, but in which in the course of the disease only dissociative signs and symptoms are displayed and therefore they see the patient as a schizophrenic right from the start. We have hardly any quarrel with this; nor would we quarrel with the view, indeed we endorse the view, that schizophrenia is an advanced form of paranoia. These two views really amount to the same.

See note 14 above.

[29] Incidentally, Christ himself was subject to a psychiatric analysis by some psychiatrists of the twentieth century. For a review of the matter, see Albert Schweitzer's famous medical dissertation, written in 1911, and originally published in German in 1913. A. Schweitzer, *The Psychiatric Study of Jesus: Exposition and Criticism* (transl. and with an introduction by Ch. R. Joy), Beacon, Boston, Mass., 1948, 1962. The reason for these works is that Christ evidently exhibited all the signs and symptoms of the Christ complex. For our

part, we deny that the Christ complex is necessarily paranoic; it is so only if it is a paranoic integrative principle. The only reason for exonerating him of the complex, is the view that he was, indeed, the Son of God. (Even the ambiguity of his own expressions can be taken as a symptom or as the rational outcome of his predicament.) All those who claim to be Julius Caesar are similarly psychotic, with the sole exception of the genuine article. But whereas the genuine article in the latter case is unproblematic, the genuine article in the former case is. Thus it is hit by the paradoxes. To see this all the clearer, all we have to do is take, instead of Julius Caesar, a case which is doubtful both on political grounds and on psychological grounds, such as cases of certain pretenders to some thrones, especially Russian. The simple question in that case is, was there ever a genuine article?

[30] For the Christ complex, see M. Rokeach's famous study, *The Three Christs of Ypsilanti: A Psychological Study,* Knopf, New York, 1964.

See also the interesting works of J. de Tonquedoc, who is a theologian, and of G. Dumas, a psychiatrist. J. de Tonquedoc, *Les maladies nerveuses ou mentales et les manifestations diaboliques,* Beachesne, Paris, 3rd ed., 1938; G. Dumas, *Le surnaturel et les dieux d'après les maladies mentales,* Presses Universitaires de France, Paris, 1946.

[31] It should be noticed that the Pavlovian approach to psychology, as well as all its derivatives, suffers from an inability to even notice, let alone resolve, the paradoxes of paranoia. For all behaviorists of all sorts psychosis may be diagnostically distinguished, if at all, only in a patient's overall poor state of integration. It is therefore no surprise that members of this school gravitate towards an etiological demarcation of psychosis rather than a diagnostic one: they notice that the entity – psychosis in general or a specific psychosis under scrutiny – is distinct, yet neither diagnosis nor course of illness do justice to the intuitively noticed distinctness. Unless they view the feeling that psychosis is distinct as caused by a (slight) tendency to exaggerate, they are bound to distinguish it etiologically. In such a case there are two known avenues of research: the chemical and the genetic – each, incidentally, operating as an integrative principle dictating a research program (as explained in Chapters 4 to 6 below). For the genetic hypothesis see Sir Julian Huxley, Ernst Mayr, Humphry Osmond, and Abram Hoffer, 'Schizophrenia As A Genetic Morphism', *Nature* 204 (1964), 220-221, and reference there. The classic biochemical paper is, H. Osmond and J. R. Smythies, 'Schizophrenia: A New Approach', *J. Mental Sci.,* 98 (1952), 309-315. For a survey of the literature see Robert Cancrow (ed.), *The Schizophrenic Syndrome,* An Annual Review, Vol. 1, 1971, Brunner-Mazel, New York, Section 4, Biochemical Studies, pp. 347-477, containing about 10 papers and many references.

There is a sense in which the genetic or chemical theories are not in any way supported or undermined by our study: for all we know, the etiology of psychoses in general and paranoia in particular may be explained with the help of a genetic and/or a chemical hypothesis: we are not concerned here with etiology but with diagnosis. But there is a sense in which we disagree with members of these schools: they may be reductionists who wish to explain away all mental factors and explain all phenomena in biological and physical terms. And here we disagree: we see a marked difference between purely biological phenomena, such as epilepsy, mental deficiency, or drugging, and psychological phenomena, such as hysterical epilepsy, dementia, or confusion. Even if one day psychology will be reduced to biology, we will have to take care of the difference between what we now consider genuine and hysterical epilepsy.

But consider the reductionists' idea that psychoses are definable not diagnostically alone, but on the basis of etiology, clinical diagnosis, pathogenesis and of course, all taken together. In principle, we have no objection to this – see note 22 to Chapter 2 below. Yet, as long as etiology is so markedly unsuccessful, we think we should try also to have a clear clinical picture, if possible, so that etiology be related to as clear and distinct an entity as possible. It is in this connection that we wish to draw attention to an interesting fact. The behaviorists find it difficult to describe psychosis, particularly paranoia, and so they are driven to a reductionist theory in the hope that etiology makes clearer what diagnosis fails to. In particular, the genetic hypothesis implies that psychoses are as common in primitive societies as among the civilized ones. The question, are there psychoses in primitive societies is disputed – see notes 30 and 31 to Chapter 3 below – simply because diagnostic tools are so inadequate to handle the question.

Indeed, for all behaviorists, Pavlov or any of his followers, paranoia is hardly discernible: it is an odd fish among odd fish. See Diana Ovlovskaja, 'Stress and Psychiatric Reactions – A Review of Contemporary Soviet Research', in L. Levi (ed.), *Society, Stress and Disease,* vol. I, *Psychosocial Environment and Psychosomatic Disease,* Proceedings of an International Interdisciplinary Symposium held in Stockholm, April 1970, Oxford University Press, New York and Toronto, 1971, pp. 247-254, especially last page. See Chapter 7, note 18, below.

32 See R. D. Laing, *The Politics of Experience and the Bird of Paradise,* 1967.

33 See T. Szasz, *The Myth of Mental Illness,* 1961.

34 Let us repeat ourselves for the sake of clarity. When claiming – against Laing and Szasz – that paranoia is a disease, we mean to say that paranoia is primarily a medical condition. We do not mean by that, that it is a full (complete) disease – in the sense that we know the unity of its etiology, pathology and clinical course (we only know its clinical course). Indeed, for the moment we can only speak here of the psychopathology in paranoia in its most rudimentary sense: as a special form of psychic organization, the etiology and pathogenesis of which we ignore (for the time being). But this very special form of psychic organization we definitely view as a psychiatric-medical condition, nontheless. See note 39 below and also Chapter 2, note 22.

35 Lawrence S. Kubie, 'The Myths of Thomas Szasz', *Bull. Menninger Clinic* 38 (1974), 497-502; p. 498: "... When he [Szasz] states there is no such thing as mental illness, he is regressing to the ancient pseudophilosophical hoax of a total dichotomy between mind and body, the psychological parallelism of the eighteenth and nineteenth centuries, that ancient ideology that made the youthful Pavlov feel that even to investigate psychological processing was heresy, an invasion of the sanctity of the soul. (Note that Pavlov in his mature years abandoned this position entirely, – Kubie, 1941, 1959.)" The reference is the author's own review of Y. P. Frolov, 'Pavlov and His School', in *Psychoanal. Quart.* 10 (1941), 329-339, and to his 'Pavlov, Freud and Soviet Psychiatry', *Behav. Sci.* 4 (1959), 29-34.

Needless to say, this criticism is wide of the mark: Szasz does not oppose any psychotherapy, only the therapist's endorsement of the moral dependence of the patient on him. See T. S. Szasz' *The Ethics of Psychoanalysis; The Theory and Method of Autonomous Psychotherapy,* Basic Books, N.Y., 1965, Delta Paperback 1969, preface, final chapter, epilogue, *et passim.* Nevertheless, as we say in the text, we feel that Szasz is not sharp enough, at least not in his *Myth of Mental Illness,* (Hoeber-Harper, New York, 1961), in saying what we think he says: mental illness has today two components, mental and moral, and he accepts only the mental and rejects the moral and, moreover, observes that without the moral component the mental component looks very different. This we fully and enthusiastically endorse and we think it comes across more clearly in his *The Ethics of Psychoanalysis* than in *The Myth of Mental Illness,* except that the former seems to speak exclusively of neurotics, not of psychotics – which is a pity.

[36] See Chapter 8, p. 79 below, where we note that in unusual cases the patient may approach a physician, but never complain about his paranoia proper.

As to the paradoxical aspect see E. Bleuler, who says about paranoia the following: "Paranoia: The construction from false premises of a logically developed and in its various parts logically connected, unshakable delusional system without any demonstrable disturbance affecting any of the other mental functions, and therefore also without any symptoms of 'deterioration', if one ignores the paranoic's complete lack of insight into his own delusional system". E. Bleuler, 1911, *Dementia Praecox or the Group of Schizophrenias,* International Universities Press, New York, 1950; p. 11.

One thing comes clearly out of this definition: The pathological characteristics of the paranoic are "the false premises", as well as the "lack of insight into ... (the) delusional system". Well, then, premises may be true or false; who decides? And even if the premises are false (and let us immediately agree that we, too, consider some premises as unquestionably false), how do they become a medical disease? Moreover, if these false premises are rooted – as Bleuler most probably thought – in the patient's affective (Freudian?) troubles, how does this fact become a medical disease? In short, we wish to raise two points here: one, that Bleuler overlooked the paradoxical nature (status) we wish to attribute to paranoia; second, the antipsychiatric point: is a psychological disturbance a medical disease?

[37] Bertrand Russell has noted that the falsehood of a belief cannot be the mark of insanity or else most people would be mad. Half seriously he offered – in his (1931) *The Scientific Outlook,* Norton, New York, 1959; Chapter X, p. 180 – a different idea to mark "the difference between sane and insane beliefs. Sane beliefs are those inspired by desires which agree with the desires of other men; insane beliefs are those inspired by desires which conflict with those of other men. We should all like to be Julius Caesar, but we recognize that if one is Julius Caesar, another is not; therefore the man who thinks he is Julius Caesar annoys us and we regard him as mad. We should all like to be immortal, but one man's immortality does not conflict with another's, therefore the man who thinks he is immortal is not mad. Delusions are those opinions which fail to make the necessary social adjustments, and the purpose of psycho-analysis is to produce the social adjustments which will cause such opinions to be abandoned.

"The reader, I hope, will have felt that the above account is in some respects inadequate. However hard we may try, it is scarcely possible to escape from the metaphysical conception of 'fact'. Freud himself, for example, when he first propounded his theory of the pervasiveness of sex, was viewed with the kind of horror that is inspired by a dangerous lunatic. If social adjustment is the test of sanity, he was insane, though when his theories came to be sufficiently accepted to be a source of income, he became sane. This is obviously absurd. Those who agree with Freud must contend that there is objective truth in his theories, not merely that they are such as many people can be got to accept."

Here Russell notes that there is truth value which differs from acceptance by society, and both of which differ from sanity or insanity; we have, then, true or false views, accepted or rejected, which can be sane or insane. We have here, then, eight possibilities, and all eight of them are realized! What, then, is the difference between the sane and the insane if it is neither truth nor acceptance, or, to use the philosophical idiom, neither nature nor convention? For the philosophic dichotomy between nature and convention see Chapters 4 and 5 below.

[38] See Chapter 2, notes 7 and 5.

[39] In the present study we have tried to avoid problems of disease in general, and more so of health, including mental health and mental hygiene. We are working on a separate monograph that will have more to say on these matters, especially in connexion with Jackson's principles. Here we should only say two things that are extremely relevant, extremely obvious, problematic for almost all writers on psychoses and utterly unproblematic for our own theory.

Jackson's principles enable us to distinguish illness from defect by viewing defect as a permanent damage, i.e., as something integrated into the system, whereas illness is something the system is engaged in, a fight from a lower level intended to restore a higher level equilibrium. This raises a number of questions we cannot enter into here. Let us only mention in passing that this forces us to distinguish diseases that are repealed from those which are tolerated – say, parasites that kill or are killed from parasites that are carried by their host at small or large but tolerated cost. In this respect we can well understand Freud's view of neuroses as modes of adjustment, usually tolerated, but which at times get out of hand and then the patient must struggle. Strangely, the worst mental condition, i.e., total withdrawal, is likewise (analytically) the end of a struggle and so, on our Jacksonian view, a defect rather than an illness.

Significantly, the milder the deviation from mental health, which is the real psychopathology in everyday life, the harder it is to declare it specifically neurotic or specifically psychotic. See Bertil Gardell's classic 'Alienation and Mental Health in the Modern Industrial Environment', in L. Levi (ed.), *Society, Stress and Disease,* vol. 1, *Psychosocial Environment and Psychosomatic Disease,* Proc. of an International Interdisciplinary Symposium held in Stockholm April 1970, Oxford University Press, New York and Toronto, 1971, pp. 148-180, 153ff.

We find it gratifying that both in mild and in very severe cases the differentiation or demarcation between neurosis and psychosis loses its force, and that even in eminently clinical cases we have neurotic psychosis as clearly distinct from psychotic neurosis (a psychotic may be hysterical, and a hysteric may have a psychotic attack). All this clearly marks Freud's essential identification between the two as a good approximation but no more than that. It is time to push on.

CHAPTER 2

[1] G. Rosen, *Madness in Society,* Routledge & Kegan Paul, London, 1968, p. 94, cites an exciting passage from Xenophon's *Memorabilia,* III, 9, 6-7 Loeb Classical Library edition, (Oxford, 1923, pp. 225-226): "most men however, he [Socrates] declared, do not call those mad who err in matters that lie outside the knowledge of ordinary people: madness is the name they give to errors in matters of common knowledge. For instance, if a man imagines himself to be so tall as to stoop when he goes through the gateway in the wall, or so strong as to try and lift houses, or to perform any other feat that everybody knows to be impossible, they say he is mad. They don't think a slight error implies madness, but just as they call strong desire love, so they name a great delusion [in the original Greek, paranoia] madness [mania]."

[2] Hirsch Jacob Zimmels, *Magicians, Theologians, and Doctors: Studies in Folk Medicine and Folk-Lore as Reflected in the Rabbinical Response 12th-19th Centuries,* Goldston, London, 1952, p. 165, quoted in H. Pollack, *Jewish Folkways in Germanic Lands (1648-1806): Studies in Aspects of Daily Life,* M.I.T. Press, Cambridge, Mass., 1971, p. 309. "That mental institutions had been first established in the 18th century is indicative of a lag in providing medical facilities for the mentally sick."

[3] Pinel's theory of mental illness as natural, and as "mental alienation", anticipates our view of the paranoic as unadjusted to public opinion (regardless of its correctness or otherwise, and of the correctness of his dissent or otherwise), so that both his view of mental illness and ours of paranoia see the illness as a kind of autism. The same holds also for Hegel's view that anyone not anchored in his community is alienated, i.e., crazy. Hegel's view anticipates most forcefully Freud's view of psychopathology in everyday life; except that it is collectivist and reactionary, need one say. Its being consistently reactionary also does not allow for the rise of the paradoxes of paranoia. For Hegel, and for any other romantic thinker, even a hero whose contribution is recognized after it was publicly endorsed should be viewed as alienated, i.e., crazy, at the creative period of his severe loneliness and isolation which is, indeed, in a sense, a period of autism. See J. Agassi, 'Genius in Science', *Phil. Soc. Sci.* 5 (1975), 145-161. The most reactionary idea of Hegel has strayed to Marx's early writings, though he must soon have recognized its reactionary character since he played it down later on. Nevertheless, contemporary Western Marxist public opinion, if it exists at all, reverts to this arch-reactionary idea as to the pinnacle of Marx's achievement. One is reminded, most forcefully, of R. D. Laing's observation as to how different an idea sounds when it comes from the mouth of a respectable philosopher, than when it comes from the mouth of a patient.

A Soviet psychiatrist, G. K. Uschakov, refers explicitly to Marx when he says, "The disease schizophrenia is not the result of an external factor which is its cause. It only translates the vital activity of the same organism in other conditions, which alienate the free play of this activity." G. K. Uschakov, 'Problèmes de diagnostic de la schizophrénie chez l'adolescent', in G. Caplan and S. Lebovici (eds.), *Psychiatric Approaches to Adolescence,* Excerpta Medica, Amsterdam, 1966; pp. 165-170, p. 166.

[4] From time to time we read in the literature of medical history about mental

patients before 1789, treated by a physician who is referred to as a philanthropist. If there were such people at the time, they were the real philanthropists, of no sociological significance, as long as their actions were not accompanied by ideology which characterizes the Enlightenment. Only this brought about some hope for change. See G. Zilboorg, *A History of Medical Psychology,* Norton, New York, 1941; F. Alexander and S. Selesnick, *The History of Psychiatry,* Harper & Row, New York, 1966; R. Hunter and I. MacAlpine, *300 Years of Psychiatry, 1535-1860,* Oxford University Press, London, 1963; K. Dörner, *Bürger und Irre,* Europäische Verlag, Frankfurt am Main, 1969; G. Mora, 'The History of Psychiatry: A Cultural and Biographical Survey', *Psychoanal. Review* 52 (2), (1965), repr. in the *Intl. J. Psychiatry* 2 (1966), 335-356.

On the other hand see M. Foucault, *Madness and Civilization: A History of Insanity in the Age of Reason,* Pantheon, New York, 1965. See note 8 to Chapter 1 above.

5 Emil Kraepelin elaborated his system of psychiatric classification in the eight editions of his textbook. A 9th edition was published posthumously, by J. Lange. In almost each additional edition Kraepelin critically examined his previous suggestion of classification, modified it, re-edited and enlarged the revised edition.

The textbook was published in 1883 (1st ed.); 1887 (2nd ed.); 1889 (3rd ed.); 1893 (4th ed.); 1896 (5th ed.); 1899 (6th ed.); 1903 (7th ed.); 1909-1915 (8th ed. 4 vols.); 1927 (9th ed.). Generally most authors refer to the last and final posthumous edition (1927, 9th ed.), briefly referred to as Kraepelin-Lange.

The name of the textbook changed during the years. In 1883 it was the *Kompendium der Psychiatrie;* later it bore the title, *Psychiatrie* (1910, 8th ed., 1927, 9th ed. vol I), *Allgemeine Psychiatrie,* J. Lange, vol. II, *Klinische Psychiatrie* (Kraepelin) vols. III and IV, new impression of the 8th ed., Barth, Leipzig.

6 The dictum, no psychosis without neurosis, is discussed at length in C. Lloyd Morgan, 1926, *Life, Mind and Spirit,* Williams and Norgate, London, 1929; Lecture 1, especially p. 9, where reference is made to T. H. Huxley who endorsed or originated it, *Essays,* ii, p. 158; *cf.* i, p. 240.

7 The definition of paranoia given here as "an insidious development of a permanent unshakeable delusional system from inner causes, in which clarity and order of thinking, willing and action are completely preserved", is quoted from the 1893 edition of Kraepelin's textbook. Since then this definition remained unchanged in the further editions of Kraepelin's textbook.

The English translation here is from W. Mayer-Gross, 1950, 'Psychopathology of Delusions: History, Classification and Present State of the Problem of the Clinical Point of View,' in F. Morel (ed.), *Psychopathologie des délires,* Hermann, Paris, 1950, pp. 59-87, p. 80.

8 E. Kraepelin, from the 1893 edition of his textbook.

9 See notes 7 and 8 above.

10 Our remark concerning Kraepelin's failure to sharply distinguish between (normal) error and delusion, is not superfluous even today and our criticism of him is not meant to be taken as derogatory. Compare Kraepelin's statements with the contribution to the 1968 *Encyclopedia Britannica* article 'Paranoid Reactions,' by N. Cameron; "A chief contributing factor to this atmosphere of tension is the tendency to self-reference, i.e., to misinterpret signs of derision

and contempt, directed at the hypersensitive person. The average person is able to shrug it off, after a brief period of resentment, or to challenge it, and thus find out his error of interpretation. The person who can neither shake off his hurt feelings nor correct his mistakes of self-reference is in more than average danger of developing delusions." (vol. 17, p. 319a).

For such a view of the delusion as a mistake, see K. Jaspers, 1913-1945, *General Psychopathology,* University of Chicago Press, Chicago, 1963; Chapter I, section I, pt. 4: "To say simply that a delusion is a mistaken idea which is firmly held by the patient and which cannot be corrected gives only a superficial and incorrect answer to the problem."

[11] E. Kraepelin (1883) quoted by H. Ey in *Les délires (généralités),* cours 1953 (révu 1967), p. 5. (Our translation.)

[12] E. Kraepelin (1927), quoted by H. Ey, *loc. cit.* (Our translation.)

[13] The danger of employing circular reasoning lies at the threshold of every physician, since in medicine the point is frequently reached where the facts are interpreted in the form of *post hoc ergo propter hoc.* This danger, as well as other possible faults of reasoning in medicine, is expressed in E. Bleuler's *Das autistische-undiziplinierte Dänke* (Springer, Berlin, 1922), a classical work classified as compulsory reading in many European medical schools to the present day. See D. Rapaport, *Organization and Pathology of Thought: Selected Sources,* Columbia University Press, New York, 1950, pp. 438-450, for Bleuler's Introduction plus four brief excerpts. See Chapter 3, note 9, below.

[14] Our Chapter 4 indicates how deep-seated was the traditional view that truth can be known and error avoided, and how popular was the corollary to it, that getting stuck in error is therefore stark madness. This theme is elaborated a great deal in the writings of Popper, Bartley, and others; see the Bibliography at the end of this volume. Popper calls this theory, the theory that truth is manifest or the theory of manifest truth. We can report from personal experience that philosophers, scientists, psychiatrists, and patients, may be greatly puzzled and disturbed when their faith in the doctrine of truth as manifest is shaken.

[15] Throughout this volume we endorse Freud's insight on mental problems as results of poor self-treatment which themselves become illnesses. Now, we do not know whether it is true, of course, but wish to draw attention to the fact that it is nothing more than positive feedback, which may be regulated or explosive, and thus a neurosis (or psychosis) of a more or less normally functioning citizen or of one in dire need of help, respectively. An example of such positive feedback in medicine is oedema (local dropsy) in the legs, as a secondary sign to renal disease, itself secondary to essential hypertension. (Notice, also, that hypertension, unlike neurosis, may be viewed purely physically.) The poor blood circulation, due to the oedema, contributes in its turn to the aggravation of the hypertension disease, a situation which obviously intensifies the oedema. Here is a self perpetuating symptom, aggravating itself – and this without being neurotic.

[16] Freud was one of the strongest believers in constitution as a determining factor even in such a psychogenetic disease as neurosis. Holding this view, he is in principle the most extremist of the mechanists or organicists amongst psychoanalysts. The question, how does his chief insight integrate within his mechanistic view, if at all, is very complicated.

[17] Clinically, there is a rule differentiate a neurotic phobia from a psychotic one. A phobia is considered neurotic if the patient is aware of and complains of the unreasonableness of his fear; and psychotic if the patient finds his (for us unreasonable) fear reasonable. We accept this distinction here and carry it all the way through, as a fundamental principle of psychopathology, applicable to phobias as to one of many special cases, of no particular prominence.
[18] The view that the ego itself can be regressive, i.e., that there are structural regressions, is not unknown in psychoanalytic circles, although it will not easily be accepted by the psychoanalytic establishment. Here is a statement by an orthodox psychoanalyst, expressing in a psychoanalytic language: S. Biran, 'Versuch zur Psychopathogenese der Schizophrenie', *Psychiatria, Neurologica, Neurochirurgia (Nederlands)* 63 (1960), 252-281, summary: "The author defines his point of view by expressing himself in favour of an approach to the problem of schizophrenia which accepts the fundamental principles of psychoanalysis but deviates from the current analytical doctrine. He states that his considerations are based on a dynamic structural conception of the instincts and the ego and id as two parallel modes of activity, and on the topical structural conception of unconscious, preconscious and transconscious. The introduction of the transconscious results from a discussion of the process of regression, which explains it as a migration of psychic contents into the transconscious. Repression in the sense of partial psychic division is regarded as the formative principle of neurosis; the formative principle of psychosis is sought in repression with additional ego regression, and instinct regression is considered the formative principle of psychopathies. The psychosis theory thus elaborated maintains that the psychotic process enters each of the three topical regions; the pathogenic complex of failures remains concealed in the unconscious. Its elaboration in the id mode of activity however, enters the preconscious, and large parts of the ego are transferred to the transconscious. When ego regression occurs in addition to instinct regression, the deepest form of schizophrenia with affective decline develops. Ego regression is a bi-phasic process which may extend either as far as an assimilation of ego and id or only as far as a rationalization of the results of the id activity. When this regresssion stops at the first phase, we have a delusional psychosis as a transition to fully developed schizophrenia. This pathogenic analysis presents no conjectures as to the etiology of psychosis; it merely renders the etiological problem more concrete by defining it as the problem of the causes of regressive tendencies." See also Appendix II below.
[19] The clinical difference between hysteria and schizophrenia is so obvious that it is indeed surprising that there is no psychopathological rule in Freud's theory of distinguishing the one from the other, not to say the manifestation of the one in the form of narcissism from the other.
[20] Referring to the definition of Hysteria in the psychiatric classification, H. Claude concluded, "the problem of hysteria is not yet resolved." See H. Claude, 'Définition et nature de l'hystérie', in *Proceedings, Congr. Alien. Neurol. Lang. Franç.*, Masson, Paris, 1907, 2 vols. Indeed. See S. Follin, J. Chazaud and L. Pilon, 'Cas cliniques de psychoses hystériques', *Evolut. Psychiat.* 2 (1961), 257-286. Since hysteric psychosis is admitted, a re-evalution of the place occupied by hysteria is strongly desirable. See S. B. Guze, 'The

Validity and Significance of the Clinical Diagnosis of Hysteria (Briquet's Syndrome)', *Amer. J. Psychiat.* 132 (1975), 138-141. For a view denying hysteria any place at all in psychiatric classification (hysteria is a misdiagnosis of neurological conditions), see E. Slater, 'Diagnosis of Hysteria', *Brit. Med. J.* i (1965), 1395-1399, as well as E. Slater and E. Glithero, 'A Follow-up of Patients Diagnosed as Suffering from Hysteria', *J. Psychosomat. Res.* 9 (1965), 9-13. As an example of the revival of interest in hysteria, see J. M. Charcot, *L'hystérie,* texts from 1877 to 1890 (ed. E. Trillat), Privat, Toulouse, 1971.

[21] Freud maintained, since his study of the Schreber case, and in all his later writings, that dementia paranoides (paranoia) is an entity clinically absolutely separate from schizophrenia, not to say from neurosis. Yet he offered no explanation. For the Schreber case, see references in note 14 to Chapter 1 above.

[22] In psychiatry as in general medicine there is room for differentiation between a clinical picture and its pathological counterpart (etiology, pathogenesis). Clinically, one may differentiate conditions according to clinical criteria; symptoms, signs, course of the disease (anamnesis, actual course of the disease, further developments, etc.); whereas pathologically one differentiates conditions according to pathological criteria: findings appearing in anatomical and histological structures, biochemical aberrations, etc. Ideally, but only ideally, correct diagnosis should be based on clinical findings of which the pathology is already known.

Thus pathology in the broad sense includes etiology, i.e., cause of the disease, pathogenesis, i.e., the mechanism of its formation, and pathology in the narrow sense, i.e., what is wrong with the patient's body; it is the hidden side of the illness. The clinical picture is the overt part; it contains the diagnostic part in particular, i.e., the signs and symptoms. The study of these two is called semiology. Of course, the distinction is superficial since the same cancer can be hidden or open, or even opened for diagnostic purposes (biopsy).

Not only is the distinction between pathology and diagnosis thus superficial; an item which lies hidden to the observer's eye today may be obvious to him tomorrow, and so pass with the growth of knowledge from pathology to diagnosis.

This can be illustrated very broadly. For, we simply do not know *a priori* what is a symptom of which disease; views on such matters may radically alter with the progress of medical knowledge. To take simple examples from known developments in medicine: there were 3 clinical conditions each distinct from the other, and each was considered as an independent disease, viz. Disseminated Lupus Erythematosus (Libman-Sacks disease); Periarteritis nodosa; Scleroderma (Dermatosclerosis). The pathology of each of these clinical entities was known but no connection between them was ever thought of, until Klemperer and others found that there is a common denominator to the different pathologic findings in these 3 clinical conditions – what was affected was the connective tissue, and in a particular way, now known as Collagen disease. The 3 clinical pictures thus became now one pathological entity – and thus one disease. See P. Klemperer, 'The Concept of Collagen Disease' and 'Systemic Lupus Erythematosus: The History of a Disease', in C. Sheba (ed.), *The Proceedings of the W.H.O./U.S.C. Medical Teaching Mission to Israel,* The Ministry of Health and the Hebrew University Medical School, Jerusalem, 1951.

The story has a surprise ending. Since Collagen disease turns out to have a manifestation in the blood, and since taking minute blood samples is an accepted diagnostic tool from time immemorial, the very discovery of this new entity which led to the discovery of this manifestation enables one to complete diagnosis with ease, when formerly it had a pathogenetic part and could not be fully diagnosed at first blush.

This is not to say that the integrative process of medical knowledge always knits symptoms into diseases: it sometimes likewise splits them. There are two classic conditions, some say three, of this kind: consumption, lumbago, and perhaps also cancer. Consumption is now not only pathologically split into tuberculosis of the lungs, emphysema, and others, where emphysema itself has soon after been split, as today the diagnosis of tuberculosis and emphysema are quite easily distinguishable. Lumbago is, of course, literally pain in the back, and can be easily divided into diverse pathologies, each with its own diagnosis. As to cancer, the sixty-four dollar question is, is cancer a disease or a symptom complex? The whole contemporary cancer research science is permeated with this question.

To conclude this cautionary remark, the integrative process of medical research may at times unite symptoms, and at times split them.

What then is, for the time being, a symptom, and what is, for the time being, a disease? Since we cannot answer this question *a priori,* we need the contemporary theoretical background to the reply. The significance of a symptom, or a sign, is in that our theory tells us that it differentiates possible diseases and can be clinically spotted with relative ease (given present day theories and present day instruments). A disease, then, is the semiology, i.e. the collection of symptoms and signs, plus etiology and pathogenesis. We would like, of course, to split the course into the clinical and prognosis, where the clinical includes anamnesis and present status, and prognosis is differential according to different possible courses of treatment. This, however, is obviously problematic.

Now take a mental condition, say classic Freudian conversion paralysis. Freud claims to have discovered its mechanism – i.e., its psychological mechanism. Now, generally, a mechanism is a part of the pathology, more precisely the pathogenesis, and when somatic usually can be found only by operation, and so is not diagnostic. In psychological rather than somatic study the operation is done not by the scalpel but in conversation – the psycho-analysis. This indeed is the root of the word. Furthermore, the advancement of the science may enable the diagnostician to see the sign and thus render the mechanism a part of the diagnosis (in a process parallel to biopsy). It is thus no surprise to us that the very progress of knowledge, due to Freud and his followers, led to a confusion here between the various parts of nosology (i.e., the theory of disease), since a part of pathology unnoticingly moved to the fore and entered diagnosis. Once noticed, it will no longer confuse practitioners.

For our part, we think we have unearthed the quasi-Freudian mechanism of paranoia, which is a fixation on an integrative principle *and* the inability to think otherwise; this, in our view, moves the mechanism of paranoia from pathogenesis to diagnosis. Of the etiology and treatment, or even course, we still know no more than what the literature tells us.

Thus when we claim that our study is diagnostic we do not mean only to limit ourselves but also to make the most of existing diagnostic tools. We think that our demarcation of mental disorders into two – the first primarily affective and local neuroses as opposed to the second, primarily intellectual global psychoses, plus their combinations and borderlines – that all this makes the most out of existing diagnostic tools and clears the field.

Moreover, etiology tells us which diseases have specific causes, e.g. malaria, which not, e.g. inflammation which may be caused by any trauma plus foreign agents. When Freud speaks of traumas, he declares etiology insignificant. See F. Brüll, 'The Trauma: Theoretical Considerations', *Israel Annals of Psychiatry* 7 (1969), 69-108. (See M. Foucault, *Naissance de la clinique*, P.U.F. Paris, 1963.)

[23] Kraepelin grouped as one disease different clinical pictures, such as Kahlbaum's catatonia; Hecker's hebephrenia; Kraepelin's dementia paranoides. This grouping was done according to a prognostic criterion: the course of the disease: they all end up in dementia. In 1911, Bleuler grouped the same clinical diseases (to which he added a fourth one – the simplex form) under schizophrenia, using a (psycho)pathological criterion which he found emphasized in all of them: the specific way of distribution of primary and secondary symptoms and signs in all of them. For our own part, we view all these as symptom-complexes (probably in accord with Bleuler, certainly with Ey); and we view the defects in thought processes etc. as primary.

Freud's personal attitude concerning the treatment of psychoses was recently illuminated by the Tausk polemic. See P. Roazen, 1969, *Brother Animal, the Story of Freud and Tausk,* Penguin, Harmondsworth, 1973, and P. Roazen, 'Ethos and Authenticity in Psychoanalysis', in *The Human Context* 4 (1972), 587. Clearly, Freud hoped that Tausk would succeed in forcing the entry of psychoanalysis into psychiatric hospitals, even though most patients there are psychotics. He expected the same of Jung. See *The Freud-Jung Letters* (ed. W. McGuire), Hogarth Press and Routledge & Kegan Paul, London, 1974.

That Freud was disappointed in both is common knowledge. Had they succeeded in their attempts to psychoanalyse psychotics, it stands to reason to assume he would have treated them more kindly. On this assumption Freud punished his associates for the limitations of his theory – which, of course, is a form of self-punishment for falling short of an extremely high ego-ideal.

[24] We do not advocate the rule – Boyle's Rule – that whenever empirical evidence and theory clash, the theory should be summarily rejected. When in conflict, either has to be rejected, and it is a matter of judgment which. (See example in note 21 above.) In the case of evidence of mixtures of neuroses and psychoses we think the reason it is repeatedly dismissed is poor and becomes poorer with the accumulation of evidence. For a discussion of Boyle's Rule, see J. Agassi, 'Sensationalism', *Mind* 75 (1966), 1-24, reprinted in his *Science in Flux,* Reidel, Dordrecht and Boston, 1975.

[25] Freud first considered transference essential for catharsis, and catharsis as cure; he gave up this catharsis theory, and for no reason at all clung to the transference theory. He noticed utter failure of transference in psychosis, viewed as the cause of the failure of psychoanalytic treatment of psychosis, and explained this failure of transference as the result of the psychotic's extreme narcissism. This led him to the odd equation of all psychoses with intensely narcissistic neurosis. See next note. Needless to say, his view of psychosis as

neurosis (of a special kind) is the major target of our attack.

S. Freud, 1916-1917, *Introductory Lectures on Psychoanalysis,* Hogarth Press, London *S.E.* 15 and 16 (1953): "The patients suffering from the narcissistic neurosis produce no transference and are, therefore, inaccessible to our efforts, not to be cured by us." See the discussion of this condition in vol. 16, the 26th Lecture; see also concluding paragraph of the 16th Lecture:

"You know that psychiatric therapy has hitherto been unable to influence delusions. Can psychoanalysis do so perhaps, by reason of its insight into the mechanism of these symptoms? No, I have to tell you that it cannot; for the present, at any rate, it is just as powerless as any other therapy to heal these sufferers. It is true that we can understand what has happened to the patient; but we have no means by which we can make him understand it himself. You have heard that I could not continue the analysis of this delusion beyond the first preliminaries. Would you then maintain that analysis of such cases is undesirable because it remains fruitless? I do not think so. It is our right, yes, and our duty, to pursue our researches without respect to the immediate gain effected. The day will come, where and when we know not, when every little piece of knowledge will be converted into power, and into therapeutic power."

[26] Why did Freud not treat psychotics? P. Racamier discusses the problem, apart from theoretical reasons, in the light of the psychoanalytic concept of countertransference, thus explaining why Freud personally refused to treat psychoses. See P. C. Racamier, 'Psychothérapie psychoanalytique des psychoses', pp. 575-690, in S. Nacht (ed.), *La psychanalyse d'aujourd'hui,* Presses Universitaires de France, Paris, 1956, pp. 576-581.

Being a materialist or organicist, Bleuler considered as primary the signs stemming directly from the (cerebral) disease itself; as secondary, all the signs that testify to activity of some lower centres in the mental organization, whatever these may be. Thus the very fact that in schizophrenia thought-processes are not properly organized, is for him a primary sign, and those clinical signs of thought disturbances such as "the splitting of thought", "looseness of associations", "incoherence", "blocking" etc., all these he deemed secondary. Consequently, all the signs of schizophrenia which he considered fundamental from the clinical point of view, he considered secondary from the psychopathological point of view. This may look paradoxical, and indeed it confuses many psychiatrists. However, a quick glance at Bleuler's 1911 monograph, even of the mere Table of Contents, will corroborate us on this point. (See note 27 below.)

The breakthrough in the field of psychotherapy of psychoses is due to P. Federn, F. Fromm-Reichmann, M. Sechehaye, etc. See P. Federn, *Ego Psychology and the Psychoses,* Basic Books, New York, 1952; F. Fromm-Reichmann, *Principles of Intensive Psychotherapy,* Chicago University Press, Chicago, 1950; M. A. Sechehaye, 1947, *Symbolic Realization,* International Universities Press, New York, 1951; H. Searles, *Collected Papers on Schizophrenia and Related Subjects,* International Universities Press, New York, 1965; H. Rosenfeld, *Psychotic States: A Psychoanalytical Approach,* International Universities Press, New York, 1965.

[27] The principle of compensation in medicine says that some seemingly morbid symptoms and signs result from beneficial compensations for ills which give different or no sign and symptom. This is known as Sydenham's principle. See Chapter 1, note 18.

Jackson's principles, mentioned above on p. 4 may be viewed as a variant of Sydenham's, and in the following way. Jackson's principles claim that the secondary symptoms and signs are nothing but a 'release' due to a defect which truncates a higher level of organization. The compensation is in that at least the lower level organizing principles operate when the higher level ones are truncated. Yet this is incorrect. For whereas a compensation is initially functional and is functionally regulated, as when one kidney works harder when the other is damaged simply because more impurities pass through it, the release of the lower organizational levels is not triggered functionally but simply by the defect of the constraining higher principle. Hence some lower organizational activities thus released may serve no function or even be dysfunctional.

Along the same line one may view Bleuler's idea of the secondary symptoms and signs of schizophrenia in their relation to the primary ones as compensations. See E. Bleuler, 1911, *Dementia Praecox or the Group of Schizophrenias,* International Universities Press, New York, 1950, Section X, pp. 348-349. See also H. Ey, 'Des principes de Hughlings Jackson à la psychopathologie d'Eugène Bleuler', *Cong. Méd. Alién et Neurol. Fran.,* Genève-Lausanne, Julliet 22-27, 1946.

[28] In the introductory remarks to the chapter dedicated to schizophrenia in O. Fenichel, *The Psychoanalytic Theory of Neurosis,* Routledge & Kegan Paul, London, 1946, 1966, the author wonders "whether the diverse schizophrenic phenomena actually have anything in common" at all (p. 413). At least he will agree that, perhaps, they all have in common, some "common, specific mental mechanisms" (p. 415). Of these common mechanisms Fenichel says that "Freud succeeded in bringing them ... into consonance with his theory of neurotic symptom formation by grouping all the schizophrenic phenomena around the basic concept of regression" (p. 415).

Fenichel stresses that "with such grouping, no judgment was given as to the somatogenic organogenesis or psychogenic origin of this regression" (p. 415). In other words, and we agree, before looking for the origins of a condition, one has to know what this condition is. Fenichel further thinks that "in different cases, the regression may have different causes, and a different range, but it always has the same great depth. It reaches back to much earlier times than does any regression in neurosis, specifically to the time when the ego first came into being" (p. 415).

Here we disagree with Fenichel and follow Alberta B. Szalita, 'Regression and Perception in Psychotic States', *Psychiatry* 21 (1958), 53-63, who says, p. 54, that Freud's concept of regression has to split to a (neurotic) "*movement* backward" and a (psychotic) decline to "*a lower level* of functioning". Fenichel's last quoted sentence makes it clear he views regression as always "movement backward", and this imposes on him Freud's idea of the straight line leading from the healthy to the neurotic and through the neurosis to the psychosis. What is specific, however, for the schizophrenic regression, according to the psychoanalytic school, is the depth of the regression. In other words we are here in the presence of a quantitative gradation; but of a quantitative regression only. There is no qualitative demarcation for the psychoanalytic school. See also note 34 below.)

[29] See note 36 and 37 below.

[30] See M. Klein, 1921-1945, *Contributions to Psychoanalysis,* Hogarth Press, London, 1948.

[31] H. Segal, 1964, *Introduction to the Work of Melanie Klein,* Basic Books, New York, 1964.
[32] J. O. Wisdom, 'Freud and Melanie Klein: Psychology, Ontology, and Weltanschauung', in C. Hanley and M. Lazarowitz (eds.), *Psychoanalysis and Philosophy,* International Universities Press, New York, 1970; 'A Methodological Approach to the Problems of Hysteria', *Int. J. Psychoanal.* 42 (1961), 231; 'Comparison and Development of the Psychoanalytic Theories of Melancholia', *Int. J. Psychoanal.* 43 (1962), 113-132.
[33] In order to understand Melanie Klein, one has to remember that she was a student of Karl Abraham. See K. Abraham, *Selected Papers,* Hogarth Press, London, 1927. Melanie Klein structuralized Freud's fantasy world by dividing fantasy objects into good and evil. This may easily be related to the Gestaltist view of the normal stress situation: under stress, animals and men see the world as polarized. Now, assume that a patient suffering anxiety mistakes it for (external) stress, i.e. projects his anxiety. He will, then, obviously, polarize the world in an erroneous way, feel persecuted, etc. Contrary to Klein, then, we view the anxiety as a primary defect and the polarization as merely secondary. Of course, the Kleinian psychiatrist may search for the source of the anxiety looking for a clue in the concrete instances of projection employed by his patient. Hence, the Kleinian psychiatrist will be selective in his attention to projections. Further, we think, at times the anxiety creates more than mere projections, namely structures of thought processes, and even integrative principles. These then became primary defects and sustain both anxiety and projection as secondary: the patient becomes paranoic. The transition, we think, is the paranoic breakdown and is arrived at by the patient's decision. The decision is arrived at out of emotional consideration – e.g. the patient feels exhausted – rather than intellectual, and so there is a voluntary resignation of one's reason. That is to say, voluntary in the sense of using one's freedom not voluntary in the sense of conscious deliberation. On the contrary, the patient is in no condition to deliberate properly and so acts foolishly.
[34] It is worthwhile to notice that Klein's individual, young or not, healthy or not, is constantly active, and so even his regression is not quite Freudian. It is interesting to notice that O. Fenichel states explicitly that the Freudian regression is quite passive. See O. Fenichel, 1946, *The Psychoanalytic Theory of Neurosis,* Routledge and Kegan Paul, London, 1966, p. 160, "regression happens to the ego" and *(loc. cit.)* "in regression the ego is much more passive".

Needless to say we side with Klein and consider regression a self-perpetuating state and so an active one. Indeed, patients repeatedly complain of tiredness.
[35] We repeat that we do not discuss etiology here. In particular, many writers look for the cause of the psychotic breakdown. Contrary to them we note, first, that a psychosis need not be preceded by a breakdown. Second, that there are neurotic breakdowns. Third, the breakdown itself, be it psychotic or neurotic, is itself a disease (or a symptom complex). Finally, as we say in note 22 above, we think it strange to center on etiology before we know the structure of which we try to explain the cause of, or the pathogenesis before we know the pathology.

See also note 34 above.
[36] T. Szasz, *The Myth of Mental Illness: Foundations of a Theory of Personal Conduct,* Hoeber-Harper, New York, 1961; T. Szasz, *Law, Liberty and Psy-*

chiatry, MacMillan, New York, 1963. T. Szasz, *Ideology and Insanity,* Anchor Books/Doubleday, New York, 1970.

[37] R. D. Laing, *The Divided Self,* Tavistock, London, 1960; Penguin, Harmondsworth, 1965; R. D. Laing, *The Politics of Experience and the Bird of Paradise,* Penguin, Harmondsworth, 1972.

[38] "Everything that will be said about hysteria pertains equally, in principle, to all other so-called mental illnesses and to personal conduct generally. The manifest diversity of mental illnesses for example, the differences between hysteria, obsessions, paranoia, etc. – may be regarded as analogous to the manifest diversity characterizing different languages." T. Szasz, *The Myth of Mental Illness,* Hoeber-Harper, New York, 1961; p. 9. We have a great sympathy with this, at least as to an articulation of a latent Freudian theme.

[39] See Karl R. Popper, *Objective Knowledge: An Evolutionary Approach,* Oxford University Press, Oxford, 1972; p. 285: "Man, some modern philosophers tell us, is alienated from his world: he is a stranger and afraid in a world he never made. Perhaps he is; yet so are animals, and even plants. They too were born, long ago, into . . . a world they never made . . ."

[40] See end of Chapter 6 below.

[41] The paranoic may look anti-conformist in his disregard for public opinion. This deluded even Laing. Yet his view of his own view, as if it were public, gives away the show. See Chapter 5, note 11; especially end of the note.

[42] Laing will probably emphatically protest against the following view of S. Arieti, *Interpretation of Schizophrenia,* Brunner, New York, 1955; p. 191; "The need for rationality is as powerful as the need to gratify the irrational emotions." And also (p. 191), "human beings cannot accept anything which to them seems irrational." Laing would say, on the contrary, long live irrationality *(La déraison):* better be irrational in experiencing myself and the world, than be rational and square in this mad society. But it seems that the seeming gap between Laing and Arieti is a much lesser one than seen at first glance. Indeed, to be and act irrationally, "to be mad" in Laing's sense, is in itself an integrative principle, and so the most rational thing to do. In other words, Laing does not really contest the demand for rationality; rather he unmasks the pretence of received opinions to rationality. And, of course, we have no intention to defend the received opinions: indeed, we too try to overturn them.

[43] In R. D. Laing and A. Esterson's, *Sanity, Madness and the Family,* Tavistock, London, 1946 (on p. 18 of the Penguin ed., Harmondsworth, 1964), and also in Laing's *The Politics of Experience and the Bird of Paradise,* 1967 (pp. 87-88 of the Penguin ed., Harmondsworth, 1972) there is a reference to Szasz' *Myth of Mental Illness,* (Hoeber, New York, 1961) which strongly suggests complete agreement between Laing and Szasz. We consider this unworthy of critical comment. Rather, let us observe this difference which we find very interesting.

Laing's views, together with those of Foucault, led to the even more radical view, advocated by D. Cooper, Maud Mannoni, T. Scheff, and others, often known as 'the labelling theory' which ascribes madness to nothing but the social environment. See M. Siegler and H. Osmond, *Models of Madness, Models of Medicine,* MacMillan, New York and London, 1974, p. 57, "Most ill people are not comforted to hear that the illness which they face every day is 'just a symptom of a sick society'." They report, pp. 52-58, the failure of a center for treatment based on this idea that mental illness is a social phenomenon, not a psychological one, and refer the reader to a report prepared by the California State Employees' Association, January 1972, called 'Where Have All the Patients Gone?' which one would be curious to see.

Laing views the psychotic breakdown as the dive into inner space which every original spirit has to take in order to explore and grow and maintain originality and individuality. The dive is dangerous and may cause irreversible harm. Public opinion may increase the pain or, still worse, altogether prevent the venture. Were public opinion altered, psychoses would continue and even more people would break down; but each breakdown would then be less dangerous.

By contrast, Szasz views public opinion as the very cause of the psychotic breakdown: let the public withdraw its special treatment and the whole phenomenon will disappear. The psychotic of Szasz's view resembles the infant in Adler's view, particularly the neurotic domineering one: he rules from a position of weakness, and he is both rewarded and punished for his undertaking of this role, whereas were his antics ignored and were he treated with patience and understanding, he would not be tempted to undertake his role.

44 One may look upon Laing's and Szasz's efforts – each in his own way – as humanizing mental illness, attempting to elevate it to the level of a normal illness and, even going further, denying it as an illness at all. All this is nothing but an ideology based on the anti-cultural movement of today (not so isolated a phenomenon), though as a whole Szasz is an old-fashioned liberal rationalist who fits contemporary laws much less comfortably than Laing. Just as 'deschooling society' is the idea that those in school should start to tear down society by tearing down school, so those who oppose mental institutions may be revolutionaries proposing the same tactics. Evidently Szasz is not one of them, nor even Laing.

The anti-psychiatry movement which is also anti-establishment in general, finds a corollary in the fact that psychiatry – though in itself revolutionary in relation to pre-psychiatry – is pro-establishment. E. Kraepelin says, "The great majority of these patients continue to live for years and thus create a heavy burden to the family and the State, with consequences which deeply affect our social life." (Quoted from the preface to the French edition of E. Kraepelin, 1900, *Leçons cliniques sur la démence précoce et la psychose maniaco- dépressive*. J. Postel (ed.) Privat, Toulouse, 1970, p. 14). One must in fairness remind the reader of the progressive flavour that this had against contemporary background. For Kraepelin's contribution to modern liberal criminology, see E. Kahn, 'Emil Kraepelin', *Am. J. Psychiat.* 113 (1956), 289-294.

45 An excellent literary example of how one retreats from sanity to insanity is given by Chekhov in his *Ward Number Six,* published in 1892. See A. Chekhov, *Ward Six and Other Stories* (transl. A. Dunnigan), Signet Books, New American Library, New York, 1965. Along with the profound and piquant description of the psychological development of the hero, the sociological aspects of the problem of mental illness, such as the question of norms, establishment conformism and non-conformism, forced hospitalization, etc., are also simultaneously admirably displayed.

The impact of Checkhov's *Ward Number Six* is overwhelming.

Lenin reacted thus: "When I read this story to the end I was filled with awe. I could not remain in my room and went outdoors. I felt as though I too was locked up in a ward." See B. R. Clarke, 'Checkhov's Tb', *Proc. Royal Soc. Med.,* 56 (1963), 1023-1026.

[46] It sounds strange that Szasz denies the pathology and admits to the suffering of the patient, when not only pathology is etymologically the science of suffering and the patient the sufferer, but also the very root of medicine is its art and science of relieving suffering. Medicine and pathology are the same. As long as Szasz is willing to treat a person, he admits his illness. Szasz himself stresses the fact that both psychiatrists and patients ought to know in advance what help the patient can expect. In a clear way, this makes the patient what he is. What Szasz combats, is the idea that the patient is not a responsible citizen: he thus wants us to see mental illness differently, not to deny its existence.

Nevertheless, we can say, Szasz is quite right in denying that mental illness is any more of an illness than lumbago – literally, pain in the back. This point, stressed by Ey who insisted that psychosis is a semiological but not etiological concept, is very significant. See H. Ey, 'Unity and diversity of schizophrenia: clinical and logical analysis of the concept of schizophrenia', *Am. J. Psychiat.* 115 (1959), 706-714, especially p. 713.

[47] Admittedly it is extremely important to understand the psychotic's language; but it is a mistake to think that the special structure of the psychotic's language is the sole defect which alone accounts for this particular form of expression (as Szasz claims): at least it would be questionable. Behind any language there are thought structures and thought processes, and these may well be the seat of the trouble.

Even for J. Lacan, for whom all mental activity is in language, there is a 'logical level' of language, which corresponds in ordinary description to nothing but thought processes; in reality it is nothing but a sort of calculus (J. Lacan, *Ecrits,* Seuil, Paris, 1966). Piaget discusses the evidence for logical structures preceding – and conditioning – linguistic structures. See his *Le structuralisme,* Presses Universitaires de France, Paris, 1968; (transl. *Structuralism,* Routledge & Kegan Paul, London, 1971; para. 17, pp. 92-96): "Speech depends on an at least partially structured intelligence" (p. 94). It seems, however, that "the reverse is also true; speech structures this intelligence" (p. 94). "The real problems of relationships between linguistic and logical structures have certainly not been solved" (p. 94). If "on the level of 'concrete operations,' operational structure precedes linguistic structure – it remains to be investigated – what exactly happens at the level of 'propositional operations,' where the language of children is modified so strikingly while their reasoning becomes hypothetico-deductive" (p. 96).

See next note.

[48] In a note on p. 120 of his *The Myth of Mental Illness,* Hoeber-Harper, New York, 1961, Szasz writes: "There are some evident similarities between what I have called *protolanguage* and Freud's concept of *Primary process* thinking and also between it and the paleologic of von Domarus and of Arieti. The difference between proto-language and the two latter constructs should become clear in the course of subsequent exposition of my thesis." (Italics in the original.) Protolanguage in Szasz' terminology means "proto", being the antonym of "Meta", refers to "something that is earlier or lower than something else" (p. 119). We must confess that though we gladly consider the difference

between Szasz' views and those of Freud and Arieti, he does not seem convincing enough to us, precisely from the point of view Szasz himself wants us to accept, namely that mental illness is not a disease. We simply cannot understand why, or cannot see how, Szasz distinguishes the patient's low order language and low order thought process, when he himself, in the subsequent exposition of his thesis, speaks favourably of Piaget's studies, and thus seemingly accepting from Piaget that this is not so for normal people. He seems to endorse there Piaget's opinion on the simultaneous growth of linguistic structures and logical (intellectual, cognitive) structures in (normal) children and adolescents, etc. In the course of his subsequent exposition of his thesis, Szasz quotes Piaget on this matter quite appropriately. Even more intriguing is the fact, that when suggesting his theory of mental illness as a game, Szasz continues to quote Piaget (1945), *Play, Dream and Imitation in Childhood,* Heinemann, London, 1951, where precisely the simultaneous growth of intellectual or cognitive structures and game playing (attitudes towards rules, heteronomy *versus* autonomy, etc.) is described. In spite of all this Szasz allows for the defect in the patient's playing but not for defects in the patient's thought structures. Similarly, the growth of social roles, etc., should also be, according to Piaget, closely linked with intellectual development (of the child, the adolescent, etc.). Piaget's 1945 ideas are more fully expressed in B. Inhelder and J. Piaget, 1955, *The Growth of Logical Thinking from Childhood to Adolescence,* Routledge & Kegan Paul, London, 1958, especially in Chapter 18, dealing with social behaviour. Szasz' book, *The Myth of Mental Illness,* was published after these two. Szasz differs from them without explicitly saying so while perhaps giving an impression to the contrary.

The way we wish to integrate the views of Ey and Piaget is precisely in taking seriously their ideas on different levels of integration as identical with different stages of structural development. This, however, seems dangerously close to the theory of ontogeny recapulating phylogeny.

See Chapter 1, note 25.

[49] *The Myth of Mental Illness,* Hoeber-Harper, N.Y., 1961, p. 305; see also the two preceding notes.

[50] In our opinion there is place for a strong argument for the reform of the whole of our concept of mental healths as long as that includes the horror of forced hospitalization, or even the horror of free hospitalization within what Goffman calls total institutions. But we neither endorse attacks upon people who advocated humane closed institutions in the past, nor do we endorse anti-psychiatry in its radical proposals which are doomed to rejection. As for an extreme anti-psychiatrist, see D. Cooper, *Psychiatry and Anti-Psychiatry,* Paladin, St. Albans, Herts., 1967, and D. Cooper, *The Death of the Family,* Penguin, Harmondsworth, 1973. For a no less extremist view, see the continental M. Mannoni, *L'enfant, sa 'maladie', et les autres,* Ed. du Seuil, Paris, 1967 (transl. *The Child, His 'Illness' and the Others,* Random House, New York, 1970). Anti-psychiatry, even if true to the last, may have to be implemented only gradually and even if it can be radically implemented today, this does not condemn all who disagree with it in the present and even less so in the past.

CHAPTER 3

[1] Daniel M'Naghten: "There are at least 10 variant spellings of this name," Royal Commission of Capital Punishment 1949-1953 Report 75 note 2 (Cmd. 8932); H.M. Stationery Office, London, 1953, quoted in S. Glueck, *Law and Psychiatry,* Johns Hopkins Press, Baltimore 1962; p. 43, note.

[2] "The most distressing thing about M'Naghten is that it sets a standard of rationality which all but the most extreme psychotics and drooling idiots can meet", says Judge D. L. Bazelon, *Equal Justice for the Unequal* (The Isaac Ray Award Lecture), Mimeographed, 1961; p. 8, quoted by T. Szasz, in his *Law, Liberty and Psychiatry,* Macmillan, New York, 1963; p. 96.

[3] It does not make sense to speak of whole societies inflicted with any psychosis, because, however strange an integrative system is, if it is accepted by a society it is reasonable for its members to accept it. And so a foreigner may look psychotic and then look normal when compared with his peers. Yet a whole society may be neurotic, and to diverse degrees. Phobias are developed in rigid societies quite consciously – see Bertrand Russell, 1954, 'Zahatopolk', in *Nightmares of Eminent Persons,* Allen & Unwin, London, 1960. Full societies or segments of a society are known to go hysterical periodically. Here our theory does full justice to intuitively felt distinctions, expressed by diverse writers, such as Arthur Koestler, in his *The Yogi and the Commissar,* J. Cape, London, 1945, and *The Lotus and the Robot*, Hutchinson, London, 1960.

[4] T. Szasz, in *Law, Liberty and Psychiatry,* Macmillan, New York, 1963, says that for E. Lemert, *Social Pathology: A Systematic Approach to the Theory of Sociopathic Behaviour,* McGraw-Hill, New York, 1951, there is a distinction to be made between primary and secondary deviation. "Deviation is said to be primary before it becomes a stable social role ... A great deal of deviation is occasional or situational (many people steal, drink to excess and hallucinate, but not all are criminal alcoholics and mental patients). A deviation is secondary if society casts the actor in a deviant role, and the deviant accepts and lives that role."

The role of this distinction, we suggest, is to account for the fact that a mental patient is a deviant in normal society and *vice versa.*

But, of course, a mere distinction can never overcome a genuine difficulty. And, we are afraid, Lemert adumbrates a difficulty but does not take the trouble to articulate it. The quote from Lemert says, we distinguish between the madman in the street and the one in a madhouse. Why do we? And why is Lemert not worried about the healthy man committed to a hospital and the sick man in the street but only about the sane man committed to the madhouse and the madman in the street? Is the illness of the sick person – physically or mentally sick – not sufficiently different from the health of the healthy one? This question is, of course, the same as the paradoxes of paranoia that we are trying so hard to make as clear and convincing as possible, before we attempt to solve them. See Chapter 6, note 17.

Assuming, then, that Lemert is bothered about the paradoxes of paranoia, perhaps he says the paradoxes hold well enough for the psychotic episode, which, using his terminology, is a primary deviance, but not for the chronic psychotic, who cannot stay out for long, whose case is secondary. Assuming this to be Lemert's solution to the paradoxes of paranoia, we must reject it. Not only is there ambulatory or sub-clinical paranoia – which we shall discuss in Chapter 6; there is also the very question, why has a chronic psychotic so often to be under special care? This question, we are afraid, is not sufficiently seriously treated by Lemert or even by Laing and Szasz. This is indeed our sole complaint against Laing and Szasz whom we greatly admire.

5 The only explanation we found of the non-existence of psychotic communities is the (Marxist?) view of Erich Wulff, who declares that all psychoses are peculiarly Western, indeed bourgeois. Pages 95-130 of his *Psychiatrie und Klassengesellschaft,* Athenäum, Frankfurt, 1972, are translated under the title 'Questions of Cross-Cultural Psychiatry', *Intl. J. Soc.* 5 (1975), 74-116, issue devoted to the topic *Beyond the Bourgeois Subject.* On p. 83 he mentions two more authors who share his finding to this effect. His and others' explanation in this issue is dual. First, in almost any society collectivism creates a group-ego which protects the individual, whereas in bourgeois society, due to severe competition, the individual internalizes an impossible super-ego and thus he is alienated and so may go crazy. It is amusing that by inner logic these thinkers came to the source of the Hegelian theory of alienation, namely that of the madman as mentally-alienated (Pinel) as reversed by Hegel to say that the alienated is mad. See note 3 to Chapter 2 on Pinel and Hegel and Marx, and note 16 to Chapter 5 on Hegel. Needless to say, Marx cannot be held responsible for the modern Marxists. Also, bourgeois writers will hardly take seriously views postulating group-ego, except the most reactionary ones. Yet, there is some justice in the view here criticized, and the high incidence of psychoses in the West is a fact and a disconcerting one. See next note.

M. J. Field, *Search for Security, An Ethno-Psychiatric Study of Rural Ghana,* Faber and Faber, London, 1960, is full of reports on psychoses, including paranoia (pp. 296-314), schizophrenia (pp. 315-352 and 454-464), and more. See also J.B. London, 'Psychogenic Disorder and Social Conflict Among the Zulu', in M. K. Opler (ed.), *Culture and Mental Health, Cross-Cultural Studies,* Macmillan, New York, 1959, pp. 351-369 – except that there the division between neurosis and psychosis is totally ignored. See also additional references in J. M. Murphy, 'Psychiatric Labeling in Cross-Cultural Perspective,' *Science* 191 (1976), 1019-1028.

6 When we claim that our explanation of the peculiarity of the paranoic *vis-à-vis* his community is adequate, we do not mean that it is true. We hope it is true; but at least we think it obvious that it satisfies some fairly standard adequacy criteria of explanation. In particular, we feel that our explanation is empirically testable, and we hope it will be tested. But here we should stress that our view is as adequate as the first which concedes, and takes due care of, the fact that accepting the norms of a given society as the standard of sanity there is both unavoidable and unsatisfactory. It is unavoidable since otherwise the only standard of sanity will be my Utopian dream, which is paranoic. It is unsatisfactory since it endorses the *status quo,* even though local standards are always too poor, and at times unbearably so. Standards of propriety in a rotten state, standards of knowledge in a backward culture, standards of excellence in a confused tradition, are obvious examples. But for our purposes we should draw attention to a very special example of unacceptable standards, namely those practiced in societies which put unbearable emotional pressures on their members, like Puritanical New England. (Thus there is some justice

even in the views criticized in the previous note.)

Yet, and this is significant, the emotional pressures may cause psychotic breakdowns but not mass psychoses; mass-psychosis is impossible since taking far granted local standards of rationality is, as explained above, quite unavoidable. But mass neurosis is possible, e.g. in Salem, Massachusetts.

See S. J. Fox, *Science and Justice: The Massachusetts Witchcraft Trial,* Johns Hopkins Press, Baltimore, 1968; Chapter VIII, 'The Defense of Insanity,' where the author quotes sources to prove that the witches were not insane. He compares contemporary records of diagnoses which prove that already in the 18th century psychosis was well enough distinguished. Indeed, due to Freud's influence we can see the witches as neurotic and so we find them more similar to psychotics than 17th and 18th century physicians!

Though Fox's book is very enlightened, the absence of any sociological and anthropological component is sorely noticeable, all the more because of its interdisciplinary character; the legal emphasis is welcome but no substitute for social history.

[7] M. Ginsberg, 'Anti-Semitism', pp. 196-212, in his *Reason and Unreason in Society: Essays in Sociology and Social Philosophy,* Longmans Green, London, 1948, pp. 197-8: "with the mass of judgments thus built up are interwoven many others designed to make them more coherent and systematic. This again takes place in accordance with well-known psychological tendencies. There is in many people a desire to be able to claim rational grounds for their beliefs, especially when their cherished convictions meet with reasoned opposition. In this way beliefs which may have very little rational ground are supplemented by other beliefs formed *ad hoc* and constitute with them a system extremely difficult to shake. Familiar examples of this process can be given from the history of religious beliefs, where we often find beliefs in the infallibility and complete reliability of the authorities". And elsewhere Ginsberg says, "When the circles are vicious, e.g. in prejudice, it is sensible to break them by a [simultaneous and] concerted attack at different points." See p. 184 of 'On Prejudice' in his *The Psychology of Society,* Methuen, London, 1964.

[8] The mechanism of reinforced dogmatism is quite simple. A theory which includes an explanation of the opposition that it encounters can be so viewed: if you agree with me, it is because you see the justice of my claim; if you disagree with me, you thereby confirm my prediction that I shall meet disagreement. This, however, can be corrected and the reinforcement can be given up. A theory which predicts that all neurotics will reject it, where neurosis is marked clearly and independently of responses to the theory, can well be tested.

See Chapter 5, note 2, below.

Freud, in his autobiography, expresses some unease about the success his American visit had: if the reception to him were negative he could explain it, and if his ideas were widely endorsed in Clark University he would have been delighted. Finally he fell back on his view of the Americans as by and large not serious intellectuals. See G. Stanley Hall, *Life and Confessions of a Psychologist,* Appleton, New York, 1924, p. 333 – on Freud's visit to the U.S.

[9] Ginsberg knows that there is a thin dividing line between prejudice and paranoia. But he does not ask, when does one cross the line and how do we demarcate the one who did from the one who did not. See note 7, above. It has been suggested that psychoses can be cured by a total treatment, akin to brain-

wash if not plainly brainwash. We cannot take this suggestion seriously.

[10] Ginsberg refers to a difference between the carrier of a prejudice and the paranoic, at least in one respect. "When a paranoiac accuses judges of corruption, lawyers of being in the pay of his enemies, and imagines a conspiracy to prevent him from obtaining justice, the root of the trouble does not lie in the nature of the lawyers and the judges or the moral theories underlying legal justice." (p. 210) In prejudice it does. And further: "In short, in so far as the antisemite really exhibits paranoiac traits, the explanation would have to begin by an examination of the mental history of the individual in question" (p. 210). See M. Ginsberg, 'Antisemitism,' pp. 196-212, in his *Reason and Unreason in Society: Essays in Sociology and Social Philosophy,* Longmans Green, London, 1948.

[11] It is a well-known fact that Freud was a pessimist in spite of his adherence to the philosophy of the Enlightenment *in toto* otherwise. See J. Agassi, 'Unity and Diversity in Science', pp. 463-522, in R. S. Cohen and M. W. Wartofsky (eds.), *Boston Studies in the Philosophy of Science,* vol. IV, Reidel and Humanities, Dordrecht and New York, 1969, especially pp. 467 and 500; reprinted in his *Science in Flux, Boston Studies in the Philosophy of Science,* vol. 28, Reidel, Dordrecht and Boston, 1975.

[12] The idea that perfectionism is crazy is very widespread. Plato refers to it as a divine madness, and as a known symptom. See J. A. Stewart, 1905, *The Myths of Plato,* MacMillan, London, 1970, Chapter on the 'Phaedrus Myth'. Bertrand Russell, in his *The Scientific Outlook* (1931), Norton, New York, 1962, speaks in Chapter XV of the perfectionism of scientists, and calls it a "Promethean madness".

[13] M. Ginsberg takes prejudice, e.g. racism, to be an integrative principle. (See notes 7 and 10 above.) We accept this with a correction: it is integrative, and it is defended *ad hoc* as an integrative principle would, but, following Jackson, we can and do see it as a partial one; and so indeed it is vulnerable to a total attack as Ginsberg says, i.e. from a higher level of integration. There is here a rigidity, but not necessarily a fixation, and when a fixation, it must be emotional. When the highest level is given up and the lower level becomes the total one, we see it as a total affliction, paranoia, not vulnerable to any rational criticism from a higher level prior to treatment.

[14] Samuel Johnson, *The History of Rasselas, Prince of Abissinia* (edited with an introduction by Geoffrey Tillotson and Brian Jenkins), Oxford University Press, London & New York, 1971.

See also note 5 to Chapter 4.

[15] There is the polarization of every human opinion into truth and error, which is identified with the polarization of opinions and actions based on them to the rational and the irrational (the logos and the mythos), and further with the polarization of everything social into nature and convention (physis and nomos). These polarizations are ancient and still extant, reflected, e.g., in Kraepelin and in Ginsberg, and very popular. They are the subject of J. Agassi, *Towards a Rational Philosophical Anthropology,* 1976 (see also reference in note 11 above). The conformism imposed in the name of these polarizations by the ideology of the Age of Reason or the Enlightenment, is discussed extensively in J. L. Talmon, *The Origins of Totalitarian Democracy,* London and Boston, 1952. The conformism is also known as the tyranny of reason or by a similar

name. See also end of next chapter.

[16] For more detail on the Romantic solution of the paradox of paranoia see J. Agassi, 'Genius in Science', *Phil. Soc. Sci.* 5 (1975), 145-161, especially pp. 150-152.

[17] Sociologically the mad are deviants who pay for their deviation by being deprived of certain customary and/or legal rights. Here Szasz's analysis comes handy: we punish a criminal either by recognizing his responsibility and charging him accordingly – or else by certifying him. (Crime is a severe form of social deviation, of course.) In this respect, of course, there is no difference between madman and genius. Now, admittedly, a mad genius may also be socially favoured: he can be a Nobel Prize winner of science or of literature. The question, then, is, how can we both reward and punish the same person? This, however, is by no means confined to our mad genius; we have the biblical story of Jonathan's case in *Samuel I,* 14 : 45 – where Jonathan both shows courage and (unknowingly) breaks an oath. The story of Tolstoy's last years is rather similar. The most sophisticated twist happened in the case of Ezra Pound, who was certified merely in order to commute his penalty for high treason, much in accord with Szasz's view, so that his certification was a penalty and a mark of recognition (and thus a reward of the highest order) at one and the same time! See also note 43 to Chapter 2 above.

[18] See note 8 above for the view of an unfailing justification as a reinforced dogmatism. Notice that though all paranoia is reinforced dogmatism, the converse is not true. In particular, the Romantic theory took care of social conflict. L. Coser, 1956, in his *Functions of Social Conflict,* The Free Press, New York, 1964, says of G. Simmel's *Conflict,* The Free Press, New York, 1955: "The central thesis of this essay is that conflict is a form of socialization. No group can be entirely harmonious, for it would then be devoid of process and structure. Conflicts are by no means altogether disruptive factors. Group formation is the result of both types of processes. Conflict as well as co-operation has social functions. A certain degree of conflict is an essential element in group formation and the persistence of group life." This view is a descendant of Romanticism; the Age of Reason saw no virtue in conflict and sought to eliminate it.

[19] The perceptive reader will notice that the suggestion to attack Enlightenment and Romantic philosophies with the aid of philosophies superior to them in accord with their own light fully conforms to our criticism of M. Ginsberg presented in note 13 above as well as to our Jacksonian approach. The general idea that new criteria have to constitute an improvement on older ones in the light of the older ones is presented in J. Agassi, 'Criteria for Plausible Argument', *Mind* 83 (1974), 406-416, reprinted in his *Science in Flux,* Reidel, Dordrecht and Boston, 1975.

[20] For the identification of the Enlightenment with traditional individualism and Romanticism with collectivism, see J. Agassi, 'Methodological Individualism', *Brit. J. Soc.,* 11 (1960), 244-270, and 'Institutional Individualism' *ibid.* 26 (1975), 144-155.

There is an exceptional philosophical trend concerned with both scientific progress and man's alienation; it is the trend of nineteenth century sociology, including such people as Auguste Comte, John Stuart Mill, and Karl Marx. Their view of social science is labelled by F. A. Hayek as scientism and attacked in his *The Counter Revolution of Science,* Free Press, Glencoe, 1964. K. R. Popper has attacked the idea shared by these philosophers of historical necessity in his *The Poverty of Historicism,* Harper and Row, New York, 1957. We shall say nothing on this trend in this study.

[21] See Max Weber, 1913, *Basic Concepts in Sociology* (transl. H. Secher), Citadel Press, New York, 1968; see also references in previous note.

[22] Karl Marx had views on individual deviants similar to those later developed by Durkheim, and now current in sociology. He thought that some degree of deviation does contribute to social stability. For example, criminals, who deviate from bourgeois norms, fulfill an important role in the cohesion of this same society. He did not believe that individual deviants could bring about social change; only a 'deviant' class can. This, of course, Durkheim rejected. As for Marx, see T. B. Bottomore and M. Rubel, eds., *Selected Writings in Sociology and Social Philosophy,* Pelican, Harmondsworth, 1963; as for Durkheim, see E. Durkheim, 1897, *Suicide: A Study in Sociology,* The Free Press, New York, 1951; see also next note.

[23] One can find the classical definition of deviation in Emile Durkheim, 1895, *The Rules of Sociological Method,* The Free Press, New York, 1958. Durkheim's view that deviation is impossible, that even criminals belong to the fabric of society and contribute to cohesion, can be found in his *The Divison of Labor* (1893), English translation by George Simpson, Free Press, Glencoe, 1933.

The contrary claim, that deviation is not only possible, but indeed vital (to society), is Simmel's. See G. Simmel, *Conflict,* The Free Press, New York, 1955, pp. 5-16. See note 18, above.

R. K. Merton and T. Parsons present deviation as a result of some incoherence within the social system – of contradictory sanctions to a given mode of conduct. See R. K. Merton, *Social Theory and Social Structure,* The Free Press, New York, rev. ed. 1957; T. Parsons, *The Social System,* The Free Press, New York, 1951. See also Albert K. Cohen's article: 'Deviation', in the *Encyclopedia of Social Sciences,* Macmillan, New York, 2nd ed., 1968. See also notes 18 above and 24 and 25 below.

[24] The very reason for the fact that Durkheim discusses deviance, whether crime or suicide, is that in his theory there is no real deviance: he is challenged to explain away seeming deviance. The question remains, did he succeed? See Percy C. Cohen, *Modern Social Theory,* Heinemann, London, 1968.

[25] Like Durkheim, Evans-Pritchard hardly explains in any detail his claim that deviants constitute an integral and integrating part of the system. See his *Nuer Religion,* Clarendon Press, Oxford,1956, p. 41.

Like Durkheim, Evans-Pritchard managed to convey this flimsy idea as part and parcel of a whole view of Man, as part and parcel of what he called "a closed system". We leave it to the reader to judge how similar Evans-Pritchard's closed system is to, if not even identical with, Ginsberg's prejudice. For Ginsberg see notes 7 and 10 above and compare with the following.

E. E. Evans-Pritchard, 1950, *Social Anthropology and Other Essays,* Free Press of Glencoe, 1962, writes: "My first book, *Witchcraft, Oracles and Magic among the Azande,* Clarendon Press, (Oxford, 1937), is about a Central African people. It is an attempt to make intelligible a number of beliefs, all of which are foreign to the mentality of a modern Englishman, by showing how they form a comprehensible system of thought, and how this system of thought is related to social activities, social structure, and the life of the individual" (p. 98). " . . . each bit of belief fits in with every other bit in a general mosaic of mystical thought. If in such a closed system of thought a belief is contradicted by a particular experience, this merely shows that the experience was mistaken, or inadequate, or the contradiction is accounted for by secondary elaborations or belief which provide satisfactory explanations of the apparent inconsistency. Even skepticism supports the beliefs about which it is exercised. Criticism of a particular diviner, for example, or distrust of a particular oracle or form of magic, merely enhances faith in others and the system as a whole" (p. 99). " . . . what at first sight seems no more than an absurd superstition is discovered by anthropological investigation to be the integrative principle of a system of thought and morals and to have an important role in the social structure" (p. 102). And the penultimate sentence of the book: "To sum it all up, I believe that social anthropology helps us to understand better, and in whatever place or time we meet him, that wondrous creature man" (p. 129).

[26] See D. H. Wrong, 'The oversocialized conception of man in modern sociology', *American Sociological Review,* 26 (1961), 183-193.

This is not to suggest that Dennis Wrong has closed the debate, nor is it an attempt to do justice to the functionalist school whether of anthropology or of other social sciences. Indeed, the whole of this topic deserves much more study and many a brilliant study has been devoted to it. See, e.g., I. C. Jarvie, *The Revolution in Anthropology,* Routledge, London, 1963; and Don Martindale (ed.), *Functionalism in the Social Sciences: The Strength and Limits of Functionalism in Anthropology, Economics, Political Science, and Sociology,* Monograph 5, in a series sponsored by the American Academy of Political and Social Science, Philadelphia, February 1965, reprint in R. Manners and D. Kaplan (eds.) *Theory in Anthropology,* Aldine, Chicago, 1968.

[27] E. Goffman, *Asylums: Essays on the Social Situation of Mental Patients and Other Inmates,* Garden City, New York. Doubleday, 1961; especially pp. 1-124.

T. Szasz, *Law, Liberty and Psychiatry,* Macmillan, New York, 1963, p. 54, says of the mental hospital as a total institution, as seen by Goffman: "The patient's authority is replaced by that of the psychiatrist and of the mental hospital as an institution." These kinds of institutions are "total because of their far-reaching control over the activities of the inmates". They are the "tuberculosis hospitals, prisons, prisoner of war camps, work camps, army barracks", etc.

[28] For the history of the invasion of mental homes by anthropologists, see A. H. Stanton and M. S. Schwartz, *The Mental Hospital,* Basic Books, New York, 1954; W. Caudil, F. C. Redlich, H. R. Gilmore, and E. B. Brody, 'Social Structure and Interaction Processes on a Psychiatric Ward', *Amer. J. Orthopsychiat.* 22 (1952), 314-334. See also A. Levy, *Les paradoxes de la liberté,* Ed de l'Epi, Paris, 1969.

See D. L. Rosenhan 'On Being Sane in Insane Places', *Science* 179 (1973), 250-258, and L. R. Spitzer, 'On Pseudoscience in Science, Logic in Remission and Psychiatric Diagnosis: a Critique on Rosenhan's "On Being Sane in Insane Places" ', *J. Abnorm. Psychol.* 84 (1975), 442-452.

Spitzer also discusses current diagnostic criteria for admission to American mental hospitals which he thinks are adequate and we think are horrid. Indeed, Spitzer makes us wonder whether our critique of the view defended by Rosenhan, which is presented in the text to this note, is not excessive. See also D. L. Rosenhan, 'The Contextual Nature of Psychiatric Diagnosis', *J. Abnorm. Psychol.* 84 (1975), 462-474.

[29] We hardly need evidence that it is commonly agreed that psychosis has a social component. See, however, note 32 below, quotation from Sir Aubrey Lewis to that effect. And notice, also, that Sir Aubrey rejects offhand, as we do, the proposal to define psychosis relative to each culture separately. See also note 17 to Chapter 6 below.

[30] The general idea that mass-hysteria is common but mass-psychosis is impossible is extremely common in the anthropological and sociological literature. Thus, in particular, shamans are usually declared hysterical types. Except for Ruth Benedict, who, in *Patterns of Cultures,* 1935, Chapter 7, end of Chapter, views them both as psychotic and as deviant – again in accord with the general idea here endorsed. The difficulty with Benedict's view is in the fact that in sessions shamans whip up mass-hysteria, not mass-psychosis. But this difficulty is not insurmountable.

Now, the only possible exception to the general idea here endorsed is to be found in a paper that we find too unclear to make a clear stand against. It is Th. Schwartz 'Cult and Context: The Paranoid Ethos in Melanesia', *Ethos* 1 (1973), 153-174. We cannot claim adequate understanding of Schwartz' position since he seems to say that a person whose paranoia fits a paranoic ethos is and is not mentally ill. But let us try. First, Schwartz makes a very strong and general claim: "The paranoid ethos may have been prevalent throughout the cultural evolutionary system of primitive societies. It persists as a potential, and is sporadically resurgent in modern societies" (p. 155). Since cultural evolutionism, as presented, is today passé among all anthropologists except in some (not all) Communist countries, we shall not comment on this.

Now, one major symptom of paranoia that Schwartz employs (pp. 155-6) is that due to V. E. van Gebsattel, (See our Chapter 8, note 18) according to which paranoia is the escape from uncertainty to (mock) certainty, especially to the certainty of the meanings the paranoic ascribes to some sort of events. Now we do not wish to quarrel with the view that this indeed happens to paranoics. But it also happens to non-paranoics, who find certitude in their mythologies, religions, and sciences, without for a moment qualifying as psychotics. Thus, it seems, what psychotic ethos and/or mass psychosis Schwartz finds in Melanesia we may also find in the nineteenth century world of science: we are no more willing to view the Melanesian ethos psychotic than the classical scientific ethos.

But this criticism is unjust, since mock-certainty is only one symptom studied by Schwartz. He also discusses persecution (p. 157). Now, we agree that finding hostile and benign meanings in the world is characteristically paranoic (see our Chapter 8, p. 80) and we agree that it is also characteristic of primitive cultures, or of magically oriented cultures (primitive or not), to be precise. See J. Agassi, 'The Limits of Scientific Explanation: Hempel and Evans-Pritchard', *Philosophical Forum* 1 (1969), 171-184. See also I. C. Jarvie and J. Agassi, 'The Problem of the Rationality of Magic', *Brit. J. Soc.* 18 (1967), 55-74, reprinted in B. Wilson (ed.), *Rationality,* Oxford University Press, Oxford, 1970, pp. 172-193, and J. Agassi and I. C. Jarvie, 'Magic and Rationality Again', *Brit. J. Soc.* 24 (1973), 236-245.

Now if certainty plus the projection of hostile (or benign) meanings make for the paranoic ethos, then the paranoic ethos is all too common. So be it. The question is, nonetheless, do we have any phenomenon of mass-psychosis akin to mass-hysteria? In particular, can we consider magic sessions, or revivalist-sessions, or mass-delusions (Lourdes, etc.), or any other mass phenomena exhibiting psychotic symptoms, to be psychotic episodes of all (or most) of their participants? Clearly, the generally accepted answer is negative. And we can explain why: as we shall see later, the autistic component in psychosis is of cardinal importance.

As we say, we do not know whether Schwartz claims that participants in the paranoic ethos are or are not paranoics themselves. He devotes to this pp. 167-8. He says, some are more tuned to the culture than others, and when the culture is paranoic the 'super-normal' is paranoic as well. We simply deny this on diagnostic grounds: there is nothing mentally troublesome, defective, sick, painful, in need of medical attention, etc., with one whose job is to be super-suspicious of men or of ghosts, be he a counter-intelligence chief or a mere shaman. (We can see how the paradox of paranoia leads Schwartz astray.) The 'symptom' of a shaman (and, we should add, of a counter-intelligence chief) is that "he persists in the role personality despite periods of waning group interest" (p. 167). Now, even when this is true, and even if it should count as psychopathological, then it is a case of individual paranoia, not of mass-psychosis.

We cannot do full justice to Schwartz's paper, and simply register our dissatisfaction with the analysis it offers which seems to us to deteriorate from one page to the next.

Finally, let us notice that the very same symptoms that are current in all magically minded societies and which Schwartz analyzes as psychotic, are usually treated as hysterical by other students of the same phenomenon. See, for a clear example, J. J. Groen, 'Social Change and Psychosomatic Disease', in L. Levi (ed.), *Society, Stress and Disease:* vol. 1: *Psychosocial Environment and Psychosomatic Disease,* Proc. of an International Interdisciplinary Symposium held in Stockholm, April 1970, New York and Toronto; Oxford University Press, 1971; pp. 91-109, and especially the section on Transcultural Studies in Nervous and Mental Disease, pp. 99-101. Of course, the phenomena, being magically meaningful, are both symbolic and cognitive, and so might be clessed as either neurotic or psychotic. But the autistic component of the cognitive side is missing: unlike the paranoic, the member of a magically orientated society is not trying to ignore any clash with accepted views.

31 Although we found it nowhere stated that psychosis and not neurosis has to be culturally relative (or related), we found ample application thereof. Thus, it is widely agreed that when it is expected of elderly people to see the little people, such phenomena, whether hallucinatory or not, need not be psychotic. The same also holds for idiosyncracies of all sorts which are expected of members of diverse groups or professions. On the other hand it is no problem to brand whole populations or groups neurotic, e.g. shamans, voodoo priests, and revivalists, and even mediaeval city dwellers. Now, of course, the symbolism of both neurosis and psychosis are clearly culturally bound – even if Jung is right about universal underlying principles. But it is the role of the symbol, not its peculiar shape, which is at stake here; and clearly, the psychotic

but not the neurotic, has to be judged against the standard of his community. See P. M. Yap, *Comparative Psychiatry,* Toronto University Press, London and Toronto, 1974. For Yap, all men have a common bio-psychological basis for mental illness, as well as syndromes that vary according to socio-cultural background.

For a recent case of 'mass hysteria', see P. D. Moss and C. P. McEvedy, 'An Epidemic of Overbreathing Among Schoolgirls' *Brit. Med. J.* ii (1966), 1295-1302.

[32] Sir Aubrey Lewis discusses the concepts of mental health, mental illness and social deviation, in 'Health as a Social Concept', *Brit. J. Soc.* 4 (1953), 109-124, reprint in A. Lewis, *The State of Psychiatry,* Routledge & Kegan Paul, London, 1967. He says: "Although social disapproval has obviously played a large part in deciding what shall be called social maladaptation, and as its main feature in current psychiatric usage, it cannot be accepted as a satisfactory criterion, varying as it does according to the group of people who express the disapproval." (p. 186). Another important discussant of these concepts is Lady Barbara Wootton. See B. Wootton, *Social Science and Social Pathology,* Allen & Unwin, London, 1959. An attempt at a synthesis is provided by D. Mechanic, *Medical Sociology: a Selective View,* Free Press, New York, 1968.

CHAPTER 4

[1] See J. Robert Oppenheimer, *Science and the Common Understanding,* B.B.C. Reith Lectures, Oxford University Press, London, 1953, end of first and of last lectures.

See also Harvey Brooks, 'Future Needs for the Support of Basic Research', in A. Lakoff (ed.), *Knowledge and Power: Essays on Science and Government,* The Free Press, New York & London, 1966; pp. 432-468, for a thorough and thought-provoking analysis which takes it for granted that support judiciously applied increases knowledge. For a critique of this view see K. Popper, *Objective Knowledge,* Clarendon, Oxford, 1972, Appendix; and G. Tullock, *The Organization of Inquiry,* Duke University Press, Durham, N.C., 1966, p. 201. Tullock mentions, as a representative of the criticized attitude, a publication by the National Academy of Science, called *Basic Research and National Goals,* 1965.

[2] Algorithms or algorisms, by definitions, are mechanical methods which ensure success. Is it possible to ensure success without a mechanical method? It seems that Oppenheimer spoke of such a guarantee without a mechanical method. And if there is a guarantee, with only high probability but not with certainty, then there might be no known mechanical procedure towards success. Yet, the high probability means that there exist a few methods which employed simultaneously, give a high relative frequency of success. See J. Agassi, 'Sensationalism', *Mind* 75 (1966), 1-24, reprinted in his *Science in Flux,* Reidel, Dordrecht and Boston, 1975.

[3] See our discussion of the public character of paranoia above and below. As for Bacon's hysterical expressions, see for example his assertion that, as Livy said of Alexander the Great "the courage to despise vain apprehension" is a necessary and sufficient condition for greatness – and one he amply fulfilled. See his *Novum Organum, Book I,* Aphorism 97. Bacon's well-known repeated

expressions of egomania are certainly not paranoic, even though he did disregard public opinion, even the very best, as quite irrational.

[4] This is the conclusion to Bacon's *Sylva Sylvarum:* some people want to conquer kingdoms of the flesh, some of the mind. The latter establish schools of thought. Doubtlessly this has become a part of our ethos and is exhibited in sophomoric debates of all sorts.

[5] Bacon stated in many places that even the most tentative hypothesis is dangerous because its upholder will, in testing it, verify it spuriously in an act of self-deception and so tentativity will soon vanish. It is intriguing to notice that Bacon declared any error, however small, may become a monomania; he gave an example from the work of William Gilbert, the famous author of *On Magnets,* 1600, who "built the whole world on a magnet".

In one place (*Novum Organum, Book II,* Aphorism 9) Bacon said the opposite: truth emerges quicker from error than from confusion. Robert Leslie Ellis, the best interpreter and the 19th century editor of Bacon's *Works,* has protested that this is a complete reversal of all that Bacon has taught and so should not represent Bacon's philosophy. Yet this is still one of Bacon's most famous aphorisms: evidently it is very useful.

[6] As to Bacon's view on scientific metaphysics, see J. Agassi, 'Unity and Diversity in Science', in R. S. Cohen and M. W. Wartofsky (eds.), *Boston Studies in the Philosophy of Science,* vol. IV, Reidel and Humanities, Dordrecht and New York, 1969; pp. 463-522, reprinted in his *Science in Flux,* Reidel, Dordrecht and Boston, 1975.

[7] A special case is Faraday's heretic views which were not discussed by his peers. The psychological impact of this on Faraday's mental health is discussed in detail in J. Agassi, *Faraday as a Natural Philosopher,* University of Chicago Press, Chicago and London, 1971.

[8] A. Einstein, 'Geometry and Experience', in his *Ideas and Opinions,* Crown, New York, 1951; p. 233, quoted also in K. Popper, *Logic of Scientific Discovery,* Hutchinson and Basic Books, London and New York, 1959; p. 314 note: "As far as the propositions of mathematics refer to reality, they are not certain; and as far as they are certain, they do not refer to reality."

We can report experiences of contacts with many cranks, especially Newtonians who view Einstein as an evil pseudo-scientist and admirers of Einstein who are convinced that the unified field theory (they seldom know much about it) will soon emerge victorious. They are very happy to use any skeptical argument against their opponents, but they may get very disconcerted when they begin to realize that perhaps this means that all certainty has now been given up as a bad job. Such cranks are not paranoic, but evidently they labor under great strain. The difference between them and sophisticated dogmatists, is that the latter are better equipped and cannot be taken by surprise, and that the latter are not as obviously laboring under the burden of great emotional pain. Which of the two is primary? Perhaps at times this, at times that; when the intellectual defect is primary we would take it as more akin to paranoia than the other way around. Will this kind of defect vanish when certainty and the desire for intellectual conquest are gone? Or will it manifest itself in a new manner?

See also note 15 below.

[9] On the success of Mill's criticism of Whewell to make philosophers disregard

him, see *Dictionary of National Biography,* Article 'Whewell, William.'

[10] Bertrand Russell (1945) writes in his *History of Western Philosophy,* Simon and Schuster, New York, 13th ed., 1967; p. 673: "It is therefore important to discover whether there is any answer to Hume within the framework of a philosophy that is wholly or mainly empirical. If not, there is no intellectual difference between sanity and insanity. The lunatic who believes that he is a poached egg is to be condemned solely on the ground that he is in a minority, or rather – since we must not assume democracy – on the ground that the government does not agree with him. This is a desperate point of view, and it must be hoped that there is some way of escaping from it." Quoted by W. W. Bartley, III, *The Retreat to Commitment,* Knopf, New York, 1962; p. 114; and by K. R. Popper, *Objective Knowledge,* Clarendon Press, Oxford, 1972, p. 5.

[11] The paradoxicality of being logical is not confined to paranoics only. See Bertrand Russell, *Religion and Science,* Butterworth, London, 1935; p. 12: "The mediaeval outlook of educated men had a logical unity which has now been lost." and p. 13: "Now logical strength is at once a strength and a weakness... The Church, in its conflict with science, exhibited both the strength and the weakness resulting from the logical coherence of its dogmas."

[12] The bibliography concerning Popper is enormous. Consult Mario Bunge (ed.), *The Critical Approach: Essays in Honor of Karl Popper,* Knopf, New York, 1964, and Paul A. Schilpp, *The Philosophy of Karl Popper,* Open Court, La Salle, Illinois, 1974, for extensive bibliographies.

[13] Michael Polanyi's *magnum opus* is his *Personal Knowledge; Toward a Post-Critical Philosophy,* Harper, New York, 1958, 1964.

See also Thomas S. Kuhn, *The Structure of Scientific Revolutions,* University of Chicago Press, Chicago and London, 1962, 1970.

[14] For the diverse senses of the word 'probability,' see K. Popper, *The Logic of Scientific Discovery,* Hutchinson, London, 1959; Appendix *IX.

[15] See J. Agassi, 'Subjectivism: From Infantile Disease to Chronic Illness', *Synthese* 30 (1975), 3-14, 33-38.

[16] The factual evidence we have in mind is not statistical – we have not counted heads – but the regular experience over twenty years, over a number of countries, learned societies, and university circles, of meeting individuals who show unquestionable signs of both paranoia and active scientific endeavor. For clarity's sake we should add that the link between the two is observed to be at times very strong, at times hardly present, at times carefully avoided. A 'mad scientist,' that is, may research his *idée fixe,* or he may be a 'split personality'.

See also note 8 above.

CHAPTER 5

[1] One way of confusing pseudo-science, metaphysics, and madness, is just to try to disregard madness at all. Indeed many treatises of logic which discuss the relationship of logic to reality say, "but this does not apply to lunatics", yet later on they denounce some pseudo-science as lunatic. This has an important practical outcome; hardly any logician has occupied himself with the logic (or perhaps logics) displayed by different patients (and in different diseases), and this is a real loss to the study of both psychopathology and of thinking. Though we have a great deal of respect for all those physicians (and

psychologists) who have attempted to study the structure of thinking in mental illness, we think it is rather clear that none of them had any measure of success in this direction.

[2] K. R. Popper, *Conjectures and Refutations: the Growth of Scientific Knowledge,* Routledge & Kegan Paul, London, 1963, pp. 49-50:

"I may perhaps mention here a point of agreement with psychoanalysis. Psycho-analysts assert that neurotics and others interpret the world in accordance with a personal set pattern which is not easily given up, and which can often be traced to early childhood. A pattern or scheme which was adopted very early in life is maintained throughout, and every new experience is interpreted in terms of it; verifying it, as it were, and contributing to its rigidity. This is a description of what I have called the dogmatic attitude, as distinct from the critical attitude, which shares with the dogmatic attitude the quick adoption of a schema of expectations – a myth, perhaps, or a conjecture or hypothesis – but which is ready to modify it, to correct it, and even to give it up. I am inclined to suggest that most neuroses may be due to a partially arrested development of the critical attitude; to an arrested rather than a natural dogmatism; to resistance to demands for the modification and adjustment of certain schematic interpretations and responses. This resistance in its turn may perhaps be explained, in some cases, as due to an injury or shock, resulting in fear and in an increased need for assurance or certainty, analogous to the way in which an injury to a limb makes us afraid to move it, so that it becomes stiff. (It might even be argued that the case of the limb is not merely analogous to the dogmatic response, but an instance of it.) The explanation of any concrete case will have to take into account the weight of the difficulties involved in making the necessary adjustments – difficulties which may be considerable, especially in a complex and changing world: we know from experiments on animals that verying degrees of neurotic behaviour may be produced at will by correspondingly varying difficulties."

[3] The influence of Bergson on the development of fringe areas in both philosophy and diverse sciences is fascinating. It is no accident that both E. Meyerson and H. Ey were confirmed Bergsonists. Piaget argued against Meyerson – whom he greatly appreciated and loved – contending that it was not identity which was the integrative principle of science, but a small number of intellectual operations which constitute sets of transformations. Identity is the end-result of such transformations. See J. Piaget, *Introduction à l'épistémologie génétique,* Presses Universitaires de France, Paris, 3 vols., 1950; vol. II, ch. V: 'Conservation et atomisme'. (See note 12 below.)

[4] K. R. Popper discusses the difference between tests in science and in technology, in his *Conjectures and Refutations,* Routledge & Kegan Paul, London, 1963; 'Three Views Concerning Human Knowledge', para. 5: 'Criticism of the Instrumentalist View', pp. 111-114. For a critique of this view see J. Agassi, 'The Confusion between Science and Technology in Standard Philosophies of Science', *Technology and Culture,* III (1966), 348-366, reprinted in his *Science in Flux,* Reidel, Dordrecht and Boston, 1975.

[5] See J. Agassi, *Faraday as a Natural Philosopher,* University of Chicago Press, Chicago & London, 1971, pp. 320-330.

[6] The significant organicist in 20th-century Europe was the influential G. de Clérambault, who was a great teacher and a first class clinician, adored and

loved by friends and foe, yet his views were hardly taken up by the next generation. His works published in the 20's present systematised delusion as a response to some kind of 'mental automatism' which operates mechanically in the patient, alarms the patient, and it is as though in his search for some ground for or sense in his fears, that the delusional system emerges, with the inner logic explaining and replacing the elements of the frightening logic in order. Clérambault's works were published posthumously in two volumes as *Oeuvre psychiatrique,* Presses Universitaires de France, Paris, 1942. For another extreme organicistic point of view, see F. Morel, *Introduction à la psychiatrie neurologique,* Masson, Paris, 1930.

7 The application of Jackson's view to mental diseases implies that its spatial and anatomical concepts have been superseded by more global and more energetic ones, where time (or space-time) replaces space, and principles of functional organization and hierarchies of functions replace Jackson's relatively simple hierarchy of merely neural functions. The application of Jackson's view to mental disease need not reduce psychology to neurology; on the contrary, the way Ey (whose approach we accept) does it, it marks their boundaries though not sharply: it shows wherein, and (much more important) how, psychology differs from and surpasses neurology.

8 See L. Pauling, 'Orthomolecular Psychiatry', *Science* 160 (1968), 265-271, (also his, 'The Molecular Basis of Genetics,' *Am. J. Psychiat.* 113 (1956), 492, quoted in L. Pauling, 'Fifty Years of Progress in Structural Chemistry and Molecular Biology,' *Daedalus* 99 (1972), 988-1014, "It is suggested that the genes responsible for abnormalities (deficiencies) in the concentration of vital substances in the brain may be responsible for increased penetrance of the postulated gene for schizophrenia, and that the so-called gene for schizophrenia may itself be a gene that leads to a localised cerebral deficiency in one or more vital substances" (p. 1013).

H. Laborit, still in molecular biology, airs another view. He writes (H. Laborit, 'Sur l'organicité moléculaire des comportements anormaux et des maladies mentales', *Agréssologie* 13 (1972), 83-91, p. 89): "For me, mental illness is only 'modulated' by the central neuromodulators, but it is stabilized on biochemical 'grounds' which are the products of the nervous system's reaction to its environment. The 'fixation' (of the illness) is organic, though produced on a molecular level of organization because of the neuronal-proteinic 'code' which is linked to the environmental experience." (Our translation.) See also F. Jacob, *La logique du vivant: une histoire de l'hérédité,* Gallimard, Paris, 1970; English transl. *The Logic of Living Systems, A History of Heredity,* Allen Lane, London, 1974.

9 No doubt organic delusions do exist as they can be chemically induced, or occur under certain known central neurological impairments. Yet even there it has not been shown that psychology is irrelevant. E.g. not all LSD takers hallucinate. Is this due to chemical peculiarities or to psychological ones? Nobody knows. Even the seemingly acausal recurrence of LSD hallucinations may be either chemically induced or, as some psychiatrists insist, due to some repressed emotional disturbances.

10 A note on the classical mind-body problem may be in order. Throughout this essay we have managed to evade it – not only in the sense that we did not discuss it or make use of any of its classic solutions. We have, following Ey,

presented and employed a generalization of Jackson's principles which transcends it: we regard the higher integrative functions as not physiological but intellectual and containing components of freedom and responsibility. Thus, our view is neither monistic nor dualistic in the traditional senses: viewed as a monism our view is peculiarly non-reductive or non-reductionist; viewed as a dualism our view is peculiar in that it fails to separate mind and body in the ordinary traditional sense. Philosophically, this is our preferred position; it is not a mere after-thought.

[11] Any definition of science is likely to idealize science not only in the sense in which Galileo's theory of gravity is an idealization that ignores air friction; it also tends to idealize in the sense that describing a society without friction idealizes it, presents it as better than it can ever be.

See J. Agassi, 'Scientists as Sleepwalkers', in Y. Elkana (ed.), *The Interaction Between Science and Philosophy,* Humanities, Atlantic Heights, 1975, pp. 391-405. See also his 'The Logic of Scientific Inquiry', *Synthese* 26 (1974), 498-514.

[12] H. Jackson was inspired by Spencer's metaphysics: H. Ey was inspired by Bergson. In other words, metaphysically, Jackson's world is materialistic and that of Ey idealistic. Both Jackson and Ey were severely attacked because of their metaphysical views; many scholastic debates took place on their differences in metaphysical outlook. Despite their general difference of views, they say the same thing. This fact leads one to believe that their metaphysics was irrelevant to their scientific views. Yet we consider their metaphysical views to be of significance; what is important is not the difference of opinion but the integrative principle common to both their metaphysics, or, more specifically, to their metaphysical theories as far as pathology is concerned, i.e. to their metapathology. Indeed, Ey is known to have said, and we have it from the horse's mouth, "a dynamic Jacksonianism" i.e. Ey's, "is the metaphysics of the illness." He also spoke of "meta-clinical" studies.

[13] As to the existentialist's insistence on commitment as more than a mere endorsement as a philosophy of life, as an answer to skepticism, see J. Agassi, 'Rationality and the *Tu Quoque* argument', *Inquiry* 16 (1973), 395-406.

[14] See J. Lacan, 'Propos sur la causalité psychique', in L. Bonnafé, H. Ey, S. Follin, J. Lacan, and J. Rouart, *Le problème de la psychogenèse des névroses et des psychoses,* Desclée de Brouwer, Paris, 1950, repr. in J. Lacan, *Ecrits,* Ed. du Seuil, Paris, 1966. He says: "What in fact is the phenomenon of delusion belief? It is, I insist, failure to recognize, with all that this term contains of an essential paradox. For to fail to recognize presupposes a recognition, as is manifested in systematic failure to recognize, where it must obviously be admitted that what is in some fashion recognized." Transl. by A. Wilden (ed.), *The Language of the Self,* Johns Hopkins University Press, Baltimore, 1968, pp. 96-97.

[15] Kant gave his celebrated courses of Lectures on 'Anthropology' every year from 1772 to 1796. He published it in 1789, as *Anthropologie in Pragmatischer Hinsicht Abgefasst*. Sect. 35-43, in Pt I, Bk I of the *Anthropologie,* has a common title, 'The Classification of Mental Disorders'. See I. Kant (1798), *Classification of Mental Disorders,* Translated and edited by C. T. Sullivan, The Doylestown Foundation, Doylestown, Pennsylvannia, 1964: "The only feature common to all mental disorders is the loss of common senses *(sensus communis),*

and the compensatory development of a unique sense *(sensus privatus)* of reasoning, e.g. a person sees in broad daylight, on his table, a light shining, which another person standing nearby does not see; or one hears a voice which no one else hears. For it is a subjectively necessary indicator of the correctness of our overall judgments, and hence of the soundness of our minds, that we compare our judgment with the judgment of others; that we do not isolate ourselves with our judgment, but on the contrary, act without private judgment as if the matter were being judged publicly" (p. vii).

"However, by the name *sensus communis* is understood the idea of a *public sense,* i.e. a critical faculty which in its reflective act takes account *(a priori)* of the mode of representation of everyone else, in order, *as it were,* to weigh its judgment with the collective reason of mankind, and thereby avoid the illusion arising from subjective and personal conditions which could readily be taken for objective, an illusion that would exert a prejudicial influence upon its judgment. This is accomplished by weighing the judgment, not so much with actual, as rather with the merely possible judgment of others, and by putting ourselves in the position of everyone else, as the results of a mere abstraction from the limitations which contingently affect our own estimate" (p. vii).

"Madness *(dementia)* is that disturbance of the mind in which everything that the madman says is indeed consistent with the formal laws of thinking, as is necessary for the possibility of an experience, but in which the subjective impressions of a falsely inventive imagination are taken for actual perceptions. Of this class are those who believe that they have enemies everywhere; who regard all the expressions, remarks, or other indifferent actions of other persons, as intended for them and as traps set for them.... Often they are, in their unfortunate madness, so ingenious in analyzing that which others unwittingly do, in order to explain it to their own satisfaction, that if their data were only correct, one would have to pay every tribute to their intelligence... I have never known anyone to recover from this disturbance (for it is a peculiar capacity to rave with intelligence). Nevertheless these individuals are not to be classed with the insane in asylums; for they, fearing only for themselves, take their supposed precautions only for their own protection, without putting others in danger; consequently they do not need to be confined for the sake of public safety. This second disturbance is methodical" (p. 15).

"As to whether there is a difference between general madness *(delirium generale)* and that which adheres fixedly to a specific object *(delirium circa objectum),* I am in doubt. Unreason (which is something positive, not just a lack of reason) is, just like reason, merely a pattern into which objects can be fitted; and both are, accordingly, based on universality. But when the insane tendency, breaking out (which usually happens suddenly), first comes into the mind's focus (thus fixing the mind upon some thought emphasized at random), then subsequently the madman raves about it more than anything else, because the first impression due to its novelty, takes a stronger hold upon him than any subsequent ones can do" (p. 18).

Hegel says that madness *(verrücktheit)* is where man claims for his singularity the validity of a universal. See G. W. F. Hegel, 1807, *Phänomenologie des Geistes,* (J. Hoffmeister's edition), Meiner, Hamburg, 1952; vol. 1, pp. 271-272. (Transl. by J. B. Baillie, *Phenomenology of Mind,* Harper Torchbook, 1967, New York, p. 397.) (Incidentally, in the same context, Hegel also speaks of the rule of fanatic priests and corrupted despots. See our Chapter 2, note 41.)

[16] For more detail see J. Agassi, *Towards a Rational Philosophical Anthropology,* forthcoming, 1976, Chapter 5.

[17] In contrast to Laing, yet while conceding much that he perceives, let us say this. One cannot avoid being deeply impressed by the commitment to ideas as experienced by many paranoics: they stand up for their convictions. They are ready to pay the price. Often we wonder whether we too would be capable of standing on our principles and paying such a high price. At the same time, the paranoic's use of private language as if it were public amounts to a wish for it to be acknowledged as public and for the public to accept it. Being a paranoic he does not succeed; but in half his mind he would if he could. This is bullying.

Both aspects of the paranoic – his commitment and his bullying – were known for some time, but only separately: we confess that we learned to see them together – and psychosis as akin to a hysterical temper tantrum – only in the course of this study. On this we think Szasz has great and profound insight.

Further, in so far as psychosis and neurosis are similar, e.g. in sharing anxiety, ambivalence, obsessiveness, the desire for acceptance, bullying, etc., there may be little difference even in their outward manifestation – obsession, dedication, tantrum, etc. Yet the neurotic mechanisms always differ from the psychotic ones, the former operating on the emotional symbolic level, and thus involving projection and introjection, the latter on the logical level. Both involve repression, but the one of true feelings, the other of true public opinion. Laing views paranoia as a defiance of public opinion; we see here great ambivalence: defiance plus acceptance!

See note 41 to Chapter 2 and the last paragraph of note 15 above.

CHAPTER 6

[1] The problem of how to class the sub-clinical or ambulant paranoic (has one to class him as a paranoic proper?) is presented here in a somewhat a similar way to M. Bleuler's discussion. See E. Bleuler, *Lehrbuch der Psychiatrie,* (ed. by M. Bleuler), Springer, Stuttgart, 1966, p. 455. He considers it a real difficulty, almost leading to a paradox (though he does not use the word paradox).

[2] Since we all gamble with life and since people under pressure are prone to flirt with risk, what at times brings a person to a breakdown is a disposition, a probability; that is to say, at times a person decides to break down or to take a collision course for a breakdown; at times, and reasonably so, he gambles with a possible breakdown or death. And so, at times, there is an element of chance built in by the ambulatory or subclinical paranoic as to whether he will become a paranoic proper. See also note 1, above.

[3] In criticism of our view of the paranoic as one who suffers, Professor F. Brüll points out the prevalent case of a young man, in his thirties, who is extremely religious, strongly believing that his mission is to bring salvation both to his people and to the world, etc. Yet he does not behave in an exaggerated manner; he is not doing more than is demanded of him by the practical rules of his religion. He does not try to convert people; he is not a burden to anyone. He does not suffer; on the contrary, he is very happy. Accidentally seen

by a psychiatrist, he was branded somewhat different from the normal. Thus, says Brüll, paranoic he is, but not a patient (etymologically: to suffer).

[4] R. D. Laing and D. G. Cooper, 1964, *Reason and Violence: A Decade of Sartre's Philosophy,* 1950-1960, (Forward by Jean Paul Sartre), Tavistock, London, 1971, present Cartesianism as bourgeois and as alienating (pp. 31-35). They advocate instead the view that "there is no pure, single individual" (p. 167). The views are evidently not only Sartre's but also of Laing and Cooper (see penultimate paragraph of the Introduction, p. 27). See notes 5 and 6 to Chapter 3.

[5] This is the concluding sentence of R. D. Laing, *The Politics of Experience and The Bird of Paradise,* Penguin, Harmondsworth, 1967, p. 156.

[6] See Joseph Fletcher, *Morals and Medicine: Moral Problems of the Patient's Right to Know the Truth, etc.,* Princeton University Press, Princeton, 1954, Beacon Press, Boston, 1960, Chapter 2, pp. 53ff; 'The Medical Code on Lying'. See also p. 43, where Fletcher quotes a doctor to say how easy it is to deceive patients, since they are disturbed to begin with. Fletcher rightly says the principles of ethics deny doctors the right to judge for their patients (pp. 37-8, 44), and that it is easy to slide from a small lie to a big one. Yet he allows (pp. 62-3) that at times it is impossible to treat *mental* patients as equals, since their very ability to understand their diagnosis is a cure (this is Freud's catharsis theory, refuted by Freud already). He even admits implicitly that even physical patients are somehow disturbed (p. 43) yet he thinks he can see the greatness of Freud's idea that there is no sharp divide between the sane and the insane – an idea fundamental to the views of both Laing and Szasz.

The idea that placebos are morally questionable is now common, but still not on the basis of the view that doctors are morally forbidden to rob patients of their responsibility even for the patients' own good and due to their ignorance. See, for a recent contribution to the literature, Sissela Bok, 'The Ethics of Giving Placebos', *Scientific American* 231 (1974), 17-23. See p. 17: "The practice is often deceptive", the author finds, "and should be restricted." See also p. 22: "The prohibition should not be absolute".

[7] H. Lasswell and R. Rubenstein, *The Sharing of Power in a Psychiatric Hospital,* Yale University Press, New Haven, 1966.

[8] A. H. Stanton and M. S. Schwartz, 'Medical Opinion and the Social Content of the Mental Hospital', *Psychiatry* 12 (1949), 243-249. See also note 28 to Chapter 3 above.

[9] E. Goffman, *Asylums: Essays on the Social Situation of Mental Patients and Other Inmates.* Doubleday, Anchor, Garden City, N.Y., 1961.

[10] We contend that a major clinical sign of all psychosis is indecision, or the inability to undertake responsibility: we find this hinted at, and in Laing and Szasz almost explicitly stated. But not quite. Let us make it further clear that the readiness to pay the price of commitment is not responsibility, whether it be religious, existentialist, stubborn, or paranoic. Responsibility is readiness to consider the price of an erroneous decision and to ask whether one can pay it. Responsibility entails the ability to choose, at times deliberately.

[11] "Today neurosis takes the place of the monasteries which used to be the refuge of all those disappointed by life or who felt too weak to face it". S. Freud, 1910, *Five Lectures on Psychoanalysis,* Hogarth Press, London, *S.E.* 11 (1975), 9-56; the 5th lecture, end of 3rd para.

[12] When applying Freud's insight about illness as a poor effort at self-treatment which is self-reinforcing, to the difference between psychosis and neurosis, especially regarding the aspects of self-awareness of the patient, it is very easy to overstep our self-imposed limitation, and hint at both etiology and suggestion of treatment. We wish to stress that we overstep only to the extent that we cannot avoid doing so.

[13] The phenomenon described here, namely the normal switching from one set to another, is called by Arthur Koestler "controlled schizophrenia" in his *The Sleepwalkers,* where he illustrates its presence in the history of scientific thought. No doubt in modern science "controlled schizophrenia" is recommended and practiced knowingly, in accord with a famous dictum of Sir William Bragg who said he thinks of quanta as waves on odd days of the week and as particles on the even ones. But we also practise it quite absent-mindedly or unwittingly. At times we dimly hope two systems of thought exist, each of which accommodates one of our mental sets and one of them accommodates the other or a third both. Yet this is only an excuse: at times we hold it yet we still keep worrying about the legitimacy of our switching back and forth between two sets (or more). Think of David Hume's bold attack on causality and his expression of a determinist view as a matter of course; think also of his intentional switching off of his philosophy and his return to common sense – which switch he effected at times by playing back-gammon. He was disturbed about the situation and made no bones about it.

[14] S. Arieti defines 'The Retreat of Reason', Chapter X, in his *Interpretation of Schizophrenia,* Brunner, New York, 1955; p. 191, in the form of a principle, the main content of which is the intellectual regression: "if, in a situation of severe anxiety, behaviour at a certain level of intellectual integration cannot take place or does not bring about the desired results, a strong tendency exists toward a behaviour of lower levels of integration in order to effect those results." The expression 'the retreat of reason' was invented by Russell and used by Popper in a political and sociological rather than psychological context, and indeed as a label for philosophical irrationalism; it is akin to Fromm's term "the escape from freedom". Arieti seems to hint at a possible psychological explanation of the phenomenon: if so, we can only say the explanation is not universally true. We shall accept Arieti's formulation, however, in the medical context. See our Chapter 9, text to note 6.

[15] As noted above, Chapter 1, notes 14 and 28, we have no objection if the reader reads schizophrenia where we write paranoia. Indeed, our chief thesis is that the two are structurally the same, though clinically they differ in their degree of organization.

[16] We do not know what to do with Laing's report of the patient's sense of distance from his own body as a symptom of schizophrenia. It is our impression that not all schizophrenics or paranoics have this idea, though, no doubt, this very idea may serve as an integrative principle for some paranoics. That it is also philosophical is, of course, paradoxical even if very few paranoics hold it.

The view that all schizophrenia is loss of a sense of the body is Federn's explanation of the dissociation which he viewed as schizophrenia; we think that dissociation is a secondary characteristic which is expressed as a diffusion of self and non-self only in accord with some 'crazy' paranoic integrative principle (on a lower level of course).

Federn's idea is complemented or echoed by Freud in his *The Ego and the Id* where he says I am both identical with my body and the master of my body. Helmut Plessner has suggested we replace Cartesian dualism with this dualism. We can hardly see the point of his idea. See R. D. Laing, *The Divided Self*, Tavistock, London, 1960: P. Federn, *Ego Psychology and the Psychoses,* Basic Books, New York, 1952; S. Freud, 1923, *The Ego and the Id,* Hogarth Press, London, *S.E.* 19 (1961), 3-66.

[17] Ideas about the normal are confused throughout the diverse literatures. The confusion is between the ideal and the average. All sorts of hybrids came out of the confusion, such as the typical member of a favoured group or a reasonable ideal (given the implementation of a reform a writer has in mind), etc. Anyway the literatures are medical, hygienic, psychological, sociological, socio-hygienic, and political. In each of these literatures there is a tendency to start with problems characteristic of certain deviants and continue with the normal until the deviant and his troubles are forgotten.

To add to the confusion, there is a moral component of whether to censure the deviant for his peculiarity, blame society for it, or let it be. Of course, when the deviancy is physical illness or handicap at times one finds it morally difficult to let it be, at times difficult not to let it be – depending on existing alternatives. Does the same hold for mental illness? (See Samuel Butler, *Erewhon.)*

There is a hot controversy as to the problem of mental illness as a social deviancy. The sociologists' labelling theory is the claim that people are referred to doctors, not because they are really mentally sick, but because they cause trouble to their society (environment). See E. Cumming and J. Cumming, *Closed Ranks,* Harvard University Press, Cambridge, 1957. The easiest way – sociologically – is for the families of deviants to label them 'sick' and thus to isolate them. Therefore, holders of the labelling theory will say, mere tolerance will drive all mental illness out of existence. (The labelling theory also holds for some physical complaints too, of course, but this is a different matter.) See T. J. Scheff, *Being Mentally Ill: a Sociological Theory,* Aldine, Chicago, 1966.

There are many psychiatrists who sympathize with the labelling theory, but not because they deny the existence of mental illness. Indeed there are two parts to this labelling theory. First, the labelling theory denies the existence of the illness does exist, this does not invalidate the sociologists' analysis of the of the person labelled mentally ill (the patient). Psychiatrists who claim that the illness exists, may still maintain that it is not necessary to isolate and thus penalize the patient. These are, then, two different issues. Moreover, even if the illness does exist, this does not invalidate the sociologists' analysis of the sociological 'unease' created by the patients not so much because of their illness and suffering, as because of the patients' complaints and request for help – because, that is, of the intolerance of their society. Therefore the question arises, do we have to contrast illness with its diagnosis as deviancy (which is the hallmark of the labelling theory)? Should one not accept from both views – the diagnosis from medical theory and the sociology from the sociological labelling theory? See A. Crowcroft, *The Psychotic,* Penguin, Harmondsworth, 1967; Introduction, pp. 11-12.

Here we see clearly not only that the norm can be either the average or the desirable, but also that the abnormal can be either sick or merely different. Thus we have two confusions reinforcing each other. Thus, possibly the deviant is sicker than the average, yet possibly healthier. For example, some clear-cut (statistical or social) deviations from the norm are nevertheless quite healthy states, such as the high – abnormal – number of red cells found in those people who live in high altitude, or the various changes found in the blood chemistry of the pregnant woman. Such changes are considered to be quite physiological and healthy – yet they are not 'normal.' The same applies to some psychiatric conditions. Indeed, for Freud, unreasonable fear is no different from its unreasonable absence – *la belle indifférence,* so called – and both he viewed as neurotic. Similarly R. S. Lazarus, and also Joseph Wolpe, have claimed that cognitive distortion occurs in stress situations: under those conditions – stress – the cognitive changes should not be considered as signs of a mental disease, but as a normal – rather, healthy – adaptive response, though a deviant one. Before one speaks of disease, therefore, one should carefully examine the question, under what circumstances does the abnormal-deviant behaviour occur? In other words, it is not the deviance in itself which equals disease; a deviancy only hints that some disease might underlie the deviant behaviour: sometimes – as in tolerable stress – there is no disease; sometimes – as in stress leading to damage – there is, especially if the damage is not reversible without treatment and/or struggle. See R. S. Lazarus, *Psychological Stress and the Coping Process,* McGraw Hill, New York, 1966. For J. Wolpe's view on the matter, see D. H. Ford and H. B. Urban, *Systems of Psychotherapy: A Comparative Study,* Wiley, New York, 1963; pp. 643-644.

It may sound trite to say that the same deviancy is healthy in some but not all conditions; but we can even say that deviancy is normal in deviant conditions. It is even trite to say that normally (but not always) a hospitalized person is ill, that it is normal for pregnant women to undergo all sorts of changes, etc. It may also be normal for old people to be weak, perhaps even demented: normal both in the statistical sense as well as in the sense of a normal (natural) phenomenon of aging, is senile dementia. Now is it a deviation? Is it a disease? We do not know.

Trite as all these points are, they raise paradoxes for the simplest reason that the word 'normal' and 'deviant' are relative (to an ensemble) but are often used as if absolute. Usually, when we take care to notice in what sense and relative to what ensemble a sample is abnormal, the paradoxes are easy to overcome. Not so in the case of the paradoxes of paranoia, and perhaps in some other paradoxes.

[18] The refusal to offer a definition of sickness should in no way be viewed as a weakness. Books on physical health which attempt to define physical sickness have similar trouble to those which attempt to define mental health. We can define tuberculosis even if we cannot define physical health, and we can say of two individuals infected with tuberculosis that the one is sick and the other is healthy and even explain why, yet without being able to offer a definition of physical health and sickness.

In order to define health we really need a whole metaphysical system of biology and of behaviour; it is not that such a system would not be of help, but that we need not wait for it before we develop our specific ideas of many specific cases.

CHAPTER 7

[1] The work of Piaget directly relevant to this chapter is especially his *Traité de logique,* Armand Colin, Paris, 1949. Piaget remarks on his indebtedness to Boole, F. Gonseth and the Bourbakies. Here we are especially interested in the joint work of Inhelder and Piaget on the *Growth of Logical Thinking,* discussion concerning the adolescent. See B. Inhelder and J. Piaget, *De la logique de l'enfant à la logique de l'adolescent: essai sur la construction des structures opératoires formelles,* Presses Universitaires de France, Paris, 1955. (English transl. *The Growth of Logical Thinking from Childhood to Adolescence,* Routledge & Kegan Paul, London, 1958; Basic Books, New York, 1958.)

[2] The theory of 'concrete thinking,' as advocated by Goldstein and others, we consider as predecessor to the one presented here. Rightly or wrongly, we modify 'concrete thinking' to 'concretized objects of abstract fixation.'

For Goldstein, see K. Goldstein and M. Scheerer, 'Abstract and Concrete Behaviour', *Psychological Monographs* 53 (1941); K. Goldstein, *Language and Language Disturbances,* Grune & Stratton, New York, 1948; K. Goldstein, 1940, *Human Nature in the Light of Psychopathology,* Harvard University Press, Cambridge, Mass., 1951.

[3] A. Ehrenzweig in his *The Hidden Order of Art: A Study in the Psychology of Artistic Imagination,* Paladin, London, 1970; p. 24: "The American psychoanalyst, Else Frenkel-Brunswik, found that certain rigid and badly integrated personalities reacted to ambiguous patterns with anxiety. This is not surprising. Like the rigid art student who cannot unfocus his attention, these people are incapable of a smooth rhythm between different levels of perception. This incapacity is due to a near-pathological dissociation of the ego functions. Because of this dissociation the untoward breakthrough of undifferentiated modes of vision threatens their rigidly focussed surface sensibilities with sudden disruption and disintegration."

The reference of Ehrenzweig is to E. Frenkel-Brunswik's 'Psychodynamics and Cognition', in R. Lindner (ed.), *Explorations in Psychoanalysis,* Julian, New York, 1953.

Ehrenzweig's book discusses throughout rigidity versus plasticity, drawing from material of diverse, and even conflicting, philosophical and psychological schools of thought. He also presses the paradoxes of paranoia hard (see Chapter 8, note 25, below) and at times in interesting manners: Artists (and we should add, scientists) at times have to disintegrate what they have so as to allow for a new integration; it is thus not degrees of integration but of plasticity that are beneficial; not degrees of disintegration but of rigidity that are troublesome. Rigidity, he shows, is or can be a complex syndrom.

For our own part, we view this syndrome as the inability to cope with too many parameters which do not always bring about anxiety, but only when it is expected and when sanctions against disappointing expectations are expected too. Also, of course, as we point out below, anxiety does lower the level of competence and so performance does become less adequate (the performer becomes able to cope with not enough parameters), leading to clinging and to increased anxiety, of course.

See also O. Fenichel, 1946, *The Psychoanalytic Theory of Neurosis,* Routledge and Kegan Paul, London, 1966, p. 578: "Often artists are afraid of losing their creative abilities if their unconscious conflicts, the source of their creativeness, were analysed. No absolute assurance can be given that an impairment of creative abilities through analysis is impossible. However, experience shows that neurotic inhibitions of creation are removed by analysis much more frequently than creativeness. Still, one must admit that in a certain minority of artists, neurosis and work seem to be so closely interwoven that it seems impossible to remove the one without impairing the other."

[4] N. Cameron broke the tradition of seeing schizophrenic thinking as a primitive form of thinking, similar to that of children. See N. Cameron, *The Psychology of Behaviour Disorders,* Houghton Mifflin, Boston, 1947. He found that schizophrenic thinking does not necessarily follow Piaget's rules, a view now reevaluated, and rightly so, we believe, by L. J. Chapman and J. P. Chapman, *Disordered Thought in Schizophrenia,* Prentice Hall, Englewood Cliffs, New Jersey, 1973, p. 216. Cameron created the concept of overinclusion, as the most characteristic feature of schizophrenic thinking. This concept opened a flow of important research of cognition in schizophrenia, which, as yet, is not concluded. See R. W. Payne, 'The Measurement and Significance of Overinclusive Thinking and Retardation in Schizophrenic Patients', pp. 77-97 in P. Hoch and J. Zubin (eds.), *Psychopathology of Schizophrenia,* Grune & Stratton, New York, 1966; B. Maher, 'The Language of Schizophrenia: a Review and Interpretation', *Brit. J. Psychiat.* 120 (1972), 3-17.

[5] It is one thing to consider regression in a descriptive sense, and altogether another one, to consider it as an explicative principle. C. Rycroft went so far as to say that "psychoanalysis is not a causal theory but a semantic one." See his 'Causes and Meaning', in C. Rycroft (ed.), 1966, *Psychoanalysis Observed,* Penguin, Harmondsworth, 1968, pp. 13-14, 17.

[6] See J. Agassi, *Science in Flux, Boston Studies on the Philosophy of Science,* vol. 28, Reidel, Dordrecht and Boston, 1975, Chapter 8: 'Towards a Theory of *ad hoc* Hypotheses'.

[7] The conspiracy theory of society, as described by Popper in his 'Towards a Rational Theory of Tradition', in his *Conjectures and Refutations,* Basic Books, New York, 1963, and elsewhere, is an interesting point in case. It is both an easy set of *ad hoc* amendments to a shaky theory and an easy source of persecutionism. It is very widespread among dogmatists, fanatics, ambulatory paranoics, and paranoics proper.

[8] For the history of phlogistonism, see J. Agassi, 1963, *Towards a Historiography of Science,* Wesleyan University Press, Middletown, Conn. 1967, Section 12.

[9] A possible counter-example for our contention that there is no mass psychosis is noted by Professor F. Brüll: there can be a mass persecutionist and even megalomanic feeling, e.g. of the persecuted members of the chosen people. No doubt, such cases do occur, but obviously, we are reluctant to call them paranoic, though they may be. We notice that centrism, so common to paranoics (egocentrism), can be found in a culture (ethnocentrism), yet not as paranoic.

[10] Note that adolescence may be viewed as a form of transitory bi-nationalism, *à la* Janusz Korczak, who viewed modern society as composed of two nations, young and adult. See his *King Matias the First.* In other words, this is why a transient psychotic state in adolescence (which has been observed more frequently than is customarily thought), would be considered by many psy-

chiatrists, in one way or another, a possible and legitimate occurrence in the natural history of adolescence. See also C. Frankenstein, *The Roots of the Ego,* Livingston, Edinburgh, 1966. According to Frankenstein, in the process of change characteristic of adolescence, the abstract replaces the concrete (e.g., friendship replaces friends). Thinking transforms the normative to be the adolescent's 'private' reality; drives become 'principles' and their satisfaction has a 'metaphysical' hold. Adolescence is not a series of successive identities (*à la* Erikson); not a bridge; structurally it is a fault. This view makes hebephrenia almost normal, as well as other psychotic episodes in adolescence. Needless to say we fully agree.

An actor may feel free to assume false identities feeling secure in his own, or passionately desirous to do so feeling he has none of his own. See Maugham's essay on Goethe, and Borges' on Shakespeare. The fact that a change of identity is at times secure, at times not, is the crux of Nigel Dennis' novel *Cards of Identity*. Note that Adler's theory is the source of the neo-Freudian identity theory as noted by Heinz L. and Rowena R. Ansbacher in their editors' preface to A. Adler, *Superiority and Social Interest,* third revised edition, Viking Press, New York, 1964; and their contribution to their *The Individual Psychology of Alfred Adler,* Basic Books, New York, 1956, Harper Torch, 1964, 1967, p. 56, "Neo-Freudian or Neo-Adlerian?"

It may be noted at once that we partly endorse R. D. Laing's theory of ontological security: normal people have it, psychotics do not; and ontological security spells political and social and intellectual conservatism and dead-end. Laing, we feel, concedes too much when he admits that ontological insecurity brings psychosis, though he is often enough right about the conservatism of ontological security. Rather, we feel, it is common-sense to claim that some flexibility and even alterability of the ego is required of a mature person in western society. Indeed, J. O. Wisdom suggests that in mature matrimonial relations the spouses throw pieces of their ego at each other: "If a new expression is wanted, I would prefer *personality-mingling*. What this consists of, I would conjecture, is a mutual *exchange* of aspects of the self. Reduced to humdrum terms, two people when their love turns to sharing, are simply like children with a ball, throwing parts of their personalities to and fro." See J. O. Wisdom, 'Freud and Melanie Klein: Psychology, Ontology and Weltanschauung,' in C. Hanley and M. Lazerowitz (eds.), *Psychoanalysis and Philosophy,* International Universities Press, New York, 1970; p. 352; italics in the original.

11 Here we see truth in Polanyi's claim that considering a system is not enough, that one has to become an apprentice of it and learn to work comfortably within it. But Polanyi concludes that this is a life-time job, and hence that one cannot really consider diverse alternatives. He thus recommends a firm identity within one system, a total commitment. We think this is obviously not a high norm for maturity. Rather, we think, a mature person is not afraid to apprentice himself for a while in different systems, schools of thought, occupations, etc., and each time make a decent job of his apprenticeship, neither afraid to go the whole way nor afraid to let go.

Thus, we must accept Evans-Pritchard's claim that one cannot switch systems easily, if ever, and this for purely intellectual reasons. (Except that he uses it to defend his religious dogmatism; and this we shall ignore here.) Hence, we will not brand as psychotic those to whom the idea has never occurred. It is the intellectual possibility plus the emotional inability – the clinging – that we are discussing. We are not the first to notice the similarity between psychosis and fanaticism; but we are the first to explain it, we think.

[12] It is agreed by most writers on critical debates, from Einstein to G. B. Shaw, that a mature interlocutor can change sides fairly well to order in a kind of mental exercise. We hardly need say that this is a traditional idea. In his autobiography Joseph Priestley ascribes his intellectual success to his training in switching sides in a debate.

[13] See E. Erikson, Preface to G. B. Blaine, Jr. and C. C. McArthur (eds.), *Emotional Problems of Students,* Appleton-Century-Croft, New York, 1961. On paranoia and adolescence, see H. S. Sullivan, *Clinical Studies in Psychiatry,* Norton, New York, 1956; pp. 154 and 156.

[14] "The genesis of a structure of the range *N* always emerges from a prior weaker structure of the range *N-1.*" J. Piaget, 'Les deux problèms principaux de l'épistémologie des sciences de l'homme', in *Logique et connaissance scientifique,* La Pléiade, Paris, 1967, p. 145.

We think that paranoia is precisely this – organizationally, a regression to such a lower range of structure. Here Piaget is a Jacksonian in the generalized sense of the word, as it appears in H. Ey, *Etudes psychiatriques,* Desclée de Brouwer, Paris, 3 vols., 1948-1954.

[15] Piaget already formulated the increasing (or decreasing) number of parameters. However, he does not refer to alternative abstract theories as we do here, in an extension of his views.

[16] Note that in this study 'primarily' is understood, all along, as both in the Jacksonian and Bleulerian sense, but definitely not in the Freudian sense.

See J. Zutt, 'Vom ästhetischen im Unterschied zum affektiven Erlebnisbereich', *Wien. Zschr. Nervenhk.* 10 (1955), 285, reprint in E. Strans and J. Zutt (eds.), *Die Wahnwelten (Endogene Psychosen),* Akademische Verlag, Frankfurt, 1963, pp. 155-168; especially p. 164ff.

"What is trust? There are people who attribute trust and suspicion to affects. Schwert calls paranoid suspicion, a mixed feeling; but generally one considers trust not an affect, but an attitude. We impart trust as we give a gift; we 'begin to believe in somebody' and then the belief is established. We may 'withdraw trust' . . .

. . . Trust and suspicion are categories of communication between people, perhaps the most important ones. But their essence is not clear. We use a pleasant sounding name here, paranoid. Etymologically this would hint at some deficit of an intellectual order. Trust, suspicion, the possibilities of being 'open' or to hide oneself, are linked. There are categories of experience here, on an aesthetic level . . .

Trust: 'The intuitive experience of belief, to be sure that I know the next one and what I expect of him.'

Suspicion: 'The intuitive experience of placing a question mark on another, wondering whether he is really frank with me or cheating me.'

Suspicion, therefore is not an affect, but an intellectual doubt. If this does happen, and one loses trust in another, then one is frightened. Seeing a revolver is not an affect; it is the significance thereof that counts. Only thereafter follows the affect." (Our translation.)

[17] Professor B. Scharfstein relates a very curious aspect of the fact that the

paranoic somewhere knows that he is deluded. Imagine that the delusion, by some stroke of luck, comes true. The paranoic's sense of reality, impaired as it is, will be totally destroyed. He will go mad in a deeper sense of despair.

[18] W. Sargant (1957), *Battle for the Mind,* Harper Perennial, New York, 1971, is an orthodox Pavlovian. He cannot properly demarcate neurosis from psychosis, need one say. (See Chapter 1, note 31.) Indeed it is surprising how insensitive he is to the difference – unless he thinks it does not exist and is shy of saying so outright. The author believes there are two and only two processes of conversion. The first is verification, which is conclusive scientific proof, and it is exceedingly rare. The second is brainwashing, mild or severe, and is exceedingly common. He is thus unable to demarcate the rational change of mind outside scientific verification from any old brainwash. For him, then, the paradoxes of paranoia do not even begin to exist, except perhaps in the case of the mad scientist. But even then he can easily resolve the paradox by compartmentalizing the scientist's diverse faculties and fields of interest, and declare him a scientist in one field and brainwashed or a brainwasher in another.

[19] This, we observe, is a criterion employed by conscientious editors and referees of learned journals, and even by conscientious men of science when consulted by colleagues. We report that they are often willing to admit inability to comprehend a new paper or monograph, and express willingness to learn what background knowledge they have missed. Only when the author of an incomprehensible paper or monograph says, or acts on the assumption, that reasonable colleagues should have little or no difficulty comprehending him, only then is he dismissed by the better colleagues who, evidently, can do nothing else. And then they usually judge him crazy.

[20] The paranoic's attitude to public opinion is a fascinating intricate matter and has a few components. The bare facts are two. First, when his delusion is contrasted with public opinion he brushes it off or even seemingly fails to hear the objection. Second, when his condition, not his delusion, is discussed, he shows awareness of his peculiarity, considers it a peculiarity, usually unfortunate but not without its positive aspects, and perhaps also, explains it away by the use of his delusion plus an *ad hoc* hypothesis, often persecutionist.

We explain this fact by the theory that the paranoic is not in the least aware of public opinion as public opinion. To make this explanation clear we should compare the paranoic's response to incredulity to his views from the stock of ideas publicly accepted to that of a scientist, whether a physicist's response to popular physical misconception or to popular myths such as the one that only a small number, usually ten or twenty, people understand Einstein, or such as the Velikovsky syndrome, or a physician responding to folk medicine, at home or abroad, or to the latest medical fad, or to Christian Science, etc. In all these cases of silly allegations, the expert may brush off the allegation at hand in response to the one who makes it, lightly or with hostility, or he may explain its folly rather condescendingly. The fact that these silly allegations are very popular only strengthens the response, especially the very expressions of superiority. The expressions of superiority may be seen in paranoics too; though, generally, with some added emotional overtone, such as marked overconfidence, nervous laughter, etc. Yet these emotional overtones are not themselves abnormal and can be seen – especially emphatic overconfidence masking nervousness – even in experienced medical practitioners of impeccable records. Of course, the more dissociated the paranoic is, the less able he will be to feign convincingly the expert's display of superiority. But this, according to our theory of dissociation as subsidary, is neither here nor there. On the contrary, we suggest, the more familiar the diagnostician is with the mannerisms of the particular group of experts whom the paranoic emulates, the easier it will be for him not to get exasperated and to locate the systematic applications of the delusion with ease – all the more so with the more dissociated patient than with the more resourceful one.

This raises the simple and intriguing question: does the paranoic, while under a specific delusion, which makes him disregard contradictory public opinion as mere childish fashion, does he at the same time seriously consider public opinion which has no direct or immediate logical bearing on his delusion? We suggest that cases indicating in the affirmative as well as in the negative exist and these may be significant for preferred courses of treatment. In any case this surely is an as yet unexplored territory. See Chapter 2, note 41.

Also, all this indicates clearly that paranoia has certain obvious culture-bound qualities, not only with respect to a given specific delusion but also with respect to Western attitudes to public opinion in general. If paranoia is indeed not observed in primitive societies, as some anthropologists report, it may well be explained by the lack of sophistication there regarding public opinion. Perhaps even the dispute may be settled by attempts to observe differences in attitude towards public opinion among primitives. (See note 29 below.) Paranoia may well then be a symptom of a society – such as ours – where sophistication tells us that public opinion is alterable, that we must nonetheless treat it as more than the mere fashion of the multitude, yet where we do not as yet have a cogent view of public opinion which the public can assimilate.

Now, empirical observation tends to agree with the hypothesis that when the degree of complexity of public opinion is higher than that of the paranoic's level of organization of thought, he will disregard public opinion completely. If the degree of complexity of the public opinion dealt with is lower – he will easily accept it as such. While we can define today, thanks to Piaget, the degree of complexity which is the level of organization of thinking (namely the number of parameters simultaneously involved; see also the same idea in Popper's methodology), there are no criteria for defining the complexity of public opinion. What we have discussed here are cases of alternatives, one private, one public, of the same level, or competing opinions. There he may lack the higher level of organization on which to compare and choose between the two, or he can, perhaps, but fails to see that since one of the two is public, its being public is itself an added parameter which is too much for him to consider within the level of his thought. Yet, we think, initially at least, it is not the level but the fixation within it that arrest his ability to add parameters, and this no doubt does have to do with his inability to give weight to public opinion as such, regardless of its being right or not. See H. G. McLaughlin, 'Psycho-logic', *Br. J. Educ. Psychol.*, 33 (1963), 61-69. On degrees of complexity in social behaviour, see T. Parsons, *The Social System*, The Free Press, New York, 1951; R. Merton, *Social Theory and Social Structure*, The Free Press, New York, rev. ed., 1957.

[21] Phenomenologically, the particular spectacles of the paranoic are a 'Gestalt-switch,' to a 'crazy world' (L. Binswanger). It might be described as follows: "The development of the paranoidal state is a result of an almost imperceptible but decisive jump of thought in the individual, which alters the perspective of his anxiety state and thus changes his outlook on the world. There is a link between the development and the jump of thoughts. It seems, and this is from clinical experience, that the jump is impossible without a latent developmental process." G. Benedetti, *Der Psychisch Leidende und seine Welt,* Hippokrates Verlag, Stuttgart, 1964; p. 128 (our translation).
[22] Without discussing the cause of mental illness, there may be some value to an allusion to a constituent of it which is common to all stress situations, normal, pre-pathological, neurotic or psychotic. It is this. Under stress we all relate everything we have our attention fixed on, to what we consider to be our distress. The result is bizarre even under the most normal conditions; yet it may lead to wild and valuable innovations even under the most abnormal conditions. Here, then, lies the root of similarity between genius, neurotic and psychotic. The genius relates his ideas to the publicly accepted world-view, and when he fails, he gives them up; the neurotic gives them a private symbolic meaning, and the psychotic creates a new world view of his own, disregarding the public one.
[23] Ludwig Wittgenstein, it is well known, said in his *Philosophical Investigation,* Blackwell, Oxford, 1953, there is no private language. Volumes were written on what this means and why he said it. That "he lived on the verge of mental illness" (N. Malcolm, *Ludwig Wittgenstein. A Memoir,* Oxford University Press, London, 1958, 1962), is, of course well-known, and so quite naturally we may assume he kept his sanity by resisting the temptation of a private language by telling himself that it does not exist. No doubt he was right: language is public, and a private language is a mere parasite on the public one (or, we say, an expression of a lower level of organization of thought).
[24] We often use the term 'logical', and always in the strict technical sense of being logically systematic or systematic in the sense of employing repeatedly one hypothesis and avoiding inconsistency, at least as much as most people do. The higher degree of systematization justifies – from a logical point of view – the paranoic's disregard of public opinion. This frightening fact is the peak of the paradoxes of paranoia. See previous few notes.
[25] Thomas S. Szasz, 'The Myth of Mental Illness' in his *Ideology and Insanity* (Anchor Books/Doubleday, New York, 1970): "Not only is there not a shred of evidence to support this idea [of the medical model of mental illness], but, on the contrary, all the evidence is the other way and supports the view that what people now call mental illnesses are, for the most part, *communications* [italics in the text] expressing unacceptable ideas, often framed in an unusual idiom" (p. 19).
[26] Hysterical academics are often prone to cling to abstract symbols, which from the abstract viewpoint seem rather poor: they are chosen for emotional reasons. Similarly, intellectuals who use overly academic terminology and mock logic to characterize schizophrenia or other psychoses do so because they are merely frightened out of their wits.
[27] The subtlety of the distinction between the abstract and the symbolic is intriguing and comes into its own in cases of paranoic hysteria and of hysteric

paranoia (not to be confused even though scarcely distinct). Thus, when we consider the most abstract concepts of science and/or religion and take their symbolic potency, we have to view the outcome of a fixation on that potency as primarily hysterical. When, on the other hand, we consider clearly a symbolic entity, concrete or abstract, and make it into an integrative principle (clever or lame, it matters very little which), we have to view the outcome of a fixation that integrative principle (of potency or whatever else) as primarily paranoic. Clearly, the hysteric fixation on the potency of an integrative principle is a possibility, and we wish to report hysterical (as we claim) fixations on such obviously integrative principles as Einstein's unified field theory, exhibited by people of all ranges of familiarity with it, from ones utterly untutored in mathematics to physicists of considerable abilities. What we think makes these cases hysterical is the fixation on potency which, as we say, we deem phallic. On the other hand, we have fixations by scientists on legitimate integrative principles, such as current scientific doctrines or, still better, scientific theories which are not dominant but also definitely not *passé*. Anyway, it is in just such cases of fixation on unorthodox views where scientists may become paranoic or border-line paranoic and tend to have illusions of grandeur about the victories they may reap when the ideas they back become established – which are dreams of potency and so hysterical. That paranoics can get hysterical at times is, of course, an observed fact. See note 8 to Appendix I.

[28] The concept of controlled schizophrenia, introduced by Arthur Koestler in his *The Sleepwalkers* becomes handy here. We wish to define controlled paranoia as an integrative be principle within clearly marked limits. For example, a scientist may be a paranoic about politics and/or about kinship but clearly sane about his professional intellectual activities; or vice versa. We wish to report having observed such cases. The fascinating thing about them is their incredible rigidity, dogmatism, or plain stupidity in the area of their fixations which is all the more outstanding when they are brilliant scientists – or, alternatively, blessed with enormous common sense. They often show other mild or severe symptoms of neuroses of all sorts, but this is another matter.

See also note 13 to Chapter 6 and note 28 to Chapter 8.

[29] Primitive cultures often project meanings into nature by the way of an integrative principle, of magic, sorcery, and the like. Yet there is no intellectual excellence and no intellectual isolation of the person holding the principle; also there is no isolation of the individual who holds principles of magic from his community at large. Quite interestingly, in a mixed society such as the Israeli society, magical beliefs are common and are scarcely psychopathological signs; yet when held by a brilliant young person it would – it does – seem to us to be a sign of hebephrenia or paranoia, mild or severe.

CHAPTER 8

[1] In the literature the word 'association' is often a synonym for 'thought process', on the hypothesis that all thinking is in one way or another a process of association in a Lockean or quasi-Lockean sense. This convention is largely due to Freud, and it remains despite the fact that associationism is now utterly discarded even by the Skinnerian school of psychology.

[2] Otto Fenichel (1946) in his *The Psychoanalytic Theory of Neurosis,* Routledge

and Kegan Paul, London, 1966, speaks of "symptoms of regression in schizophrenia" (pp. 417-424) and "restitutional symptoms in schizophrenia" (pp. 424-439). He says that the "first category of symptoms embraces phenomena such as fantasies of world destruction, physical sensations, depersonalization, delusions of grandeur, archaic ways of thinking and speaking, hebephrenic and certain catatonic symptoms. The second category embraces hallucinations, delusions, most of the schizophrenic social and speech peculiarities, and other symptoms" (p. 417).

One easily sees similarities as well as differences when comparing Fenichel's views with those of E. Bleuler. See E. Bleuler, 1911, *Dementia Praecox or the Group of Schizophrenias,* International Universities Press, New York, 1950; especially pp. 348-349.

[3] We are aware of the defect of our view in that it is not sufficiently empirically testable, and hope it will become so when extended to cover etiology and treatment. In the meantime, however, we suggest that in our very challenge of the empirical clinical data regarding paranoia and schizophrenia, as well as in our delineation of varieties of paranoia, there is some scope for empirical tests.

[4] There is a popular view that a major sign of schizophrenia, as opposed to paranoia, may be the persistence of the paranoic delusion as opposed to the on-and-off character of the schizophrenias. This is not true. The accepted clinical difference between a paranoic and a paranoid-schizophrenic lies mainly in that the paranoic acts accurately and in complete accordance with the delusion ("clarity and order of thinking... and action are... preserved"), whereas a paranoid-schizophrenic displays a 'split' between thought and action – his acts and deeds though resulting from the delusion, are not coherent which ever way we look at them, particularly, of course, when in utter inaction. For our own part, we see this incoherence as a part of the deficiency at a point of a lower level of organization.

[5] Some clinicians, especially those of the French school, admit to hallucinations in paranoia. But on the whole, this does not affect our discussion.

[6] Bertrand Russell raised, early in his career, the question, what happens when a humane person holds a cruel theory (see his *My Philosophical Development,* Allen & Unwin, London, 1959). Possibly, of course, when the conflict materializes, the person in question will change his opinion, possibly he will become cruel and inhuman; it is possible, finally, that he will be inconsistent at points. Russell notes that the first alternative is rare and he asks whether the second or the third alternative is preferable, and decides that humaneness is preferable to logic. Fanatics, of course, have contempt for this option. The more marked their contempt, the nearer their case is to that of paranoia.

[7] On reinforced dogmatism see Chapter 3, note 8.

[8] Bateson *et al* developed a theory of schizophrenia which covers etiology, diagnosis and therapy. See G. Bateson, D. D. Jackson, J. Haley, J. Weakland, 'Toward A Theory of Schizophrenia', *Behavioural Science,* 1 (1956), 251-264. They claim (p. 251) that their "approach is based on that point of communications theory which Russell has called the Theory of Logical Types" – they refer to Whitehead and Russell, *Principia Mathematica.* We are pained to observe that the *Principia* has nothing to do with communication theory. "The central thesis of this theory" say Bateson *et al.* "is that there is a discontinuity between a class and its members. The class cannot be a member of itself" etc.

(p. 251). Now the first sentence is naturally meaningless and the second artificially made so by the decree of the theory of types. See note 12 to Chapter 1 above. (We had no intention of commenting in detail on such a paper, but we were forced to – see the postscript to our preface.) So let us stop quoting in detail and offer the reader the gist of that celebrated paper.

What Bateson *et al.* contend is what quite a few psychologists have observed, and what we have repeatedly mentioned as a central item in the view propounded here: psychotics are prone to confuse the abstract with the concrete.

To this Bateson *et al.* offer a second factor: that of the "double bind", for which see next note, where a person is trapped between impossible options.

Now, consider a case in which a child is in a double bind yet forbidden from commenting on the situation (in what logicians will call the metalanguage; it is not clearly stated or even hinted anywhere, but, since the theory of types is not explicitly invoked until the final section which is titled 'Current position and future prospects', quite possibly Bateson *et al.* suggest the metalanguage to be a higher Russellian type, an error common in all introductory logic courses). No doubt, the comment would be abstract and it (as well as its inhibition) carries an enormous emotional force. No doubt this can be pathogenic and the psychopathology it may cause may be the confusion of the abstract with the concrete. This, however, is a far cry from an etiology, for etiology speaks of sufficient, or of necessary and sufficient, conditions, not of necessary ones, much less of possible ones. In one place (p. 258, right column, Summary) Bateson *et al* suggest that putting a child in a double bind and not allowing him to complain is a sufficient cause for schizophrenia. This theory is palpably false and we report that we know of many instances to the contrary.

See also next note.

[9] Here is the place to mention the theory of the double bind, so-called, which has won tremendous acclaim, which G. Bateson *et al.* presented as the etiology (as well as diagnosis) of schizophrenia (see previous note), and which is already described by Freud (1923) in his *The Ego and the Id,* Hogarth Press, London, *S.E.* 19 (1962), 3-66, as the source of the Oedipus complex. The "double bind", the "sticky wicket", or, as Freud has put is, "you are damned if you do and you are damned if you don't" is the case of only bad options, of feeling trapped, and of anxiety, perhaps also of a sense of guilt. We think that the double-bind, i.e. the feeling of being cornered, is indeed common to all problems of mental patients, neurotics or psychotics, severe or slight. It is also known in ordinary situations, as described by Kafka, Sartre, and others. (You do not have to be a paranoic to feel persecuted.)

[10] The onset of the disease is, according to our theory, post-adolescent, more likely (it being a regression) in an early phase. The signs of the disease are usually observed at the third and fourth decades of life. We suspect that the delay between onset and observation is a cultural determinant: clearly the limit to police patience which brings a patient to the clinic is not decided by the patient; the patient may, however, adjust himself so as to live around that limit. But, since he is not always so very flexible, the limit of police patience may be a major factor in the case and so be a factor determining the average age of the onset of the disease as seen by the clinician. That is to say, in the first approximation all patients start at the same age and their negative manifestations become increasingly conspicuous at the same pace, but the police of one

country may respond quicker to minor disturbances than others. In the second approximation, the patient's manifestations are geared to police patience and so they all come to the attention of the clinician at about the same age. The third approximation will combine the first two. The fourth approximation will add to the police flexibility that of the family and the employer.

[11] It is reported that women are statistically more frequent victims of the disease. If our theory is true, and if this is the case, then it may be culturally related to the intellectually inferior position of women in our culture, which leads more women to be fringe-intellectual than men, rather than anything inherent in sex-difference. It may, of course, also relate to the higher pressure on women in our society, especially on the fringe-intellectual ones.

[12] P. Sérieux and J. Capgras, in *Les folies raisonnantes: le délire d'interprétation,* Alcan, Paris, 1909, see the contradictory phenomenon of a delusion as side by side with an otherwise intact mental activity, but they do not elaborate on that further. G. Dumas, *Le surnaturel et les dieux d'après les maladie mentales,* Presses Universitaires de France, Paris, 1946, is more perceptive about this contradiction, but not enough. See note 13 below, and Chapter 1, note 13.

[13] See H. Ey, *Les délires (généralités),* cours 1953 (revu 1967).

[14] We may sound bluntly inconsistent when we declare paranoia to be both a matter of degree and a jump. We do not think we are. Every decision is a jump, but some decisions are big, some small, and one man may arrive at a given stage by one big jump, another by a series of small ones.

[15] Ego-centrism is a system with the ego at the center. Centrism is a generalization of that, first proposed by Piaget. Opposed to centrism there may be polycentrism, no system, or the alternate use of different systems. The latter is called by A. Koestler *(The Sleepwalkers),* "controlled schizophrenia."

[16] Konrad Lorenz, 'Companionship in Birds', in C. H. Schiller (ed.), *Instinctive Behaviour: The Development of a Modern Concept,* International Universities Press, New York, 1957, p. 91.

[17] Referring to K. Conrad, *Die Beginnende Schizophrenie,* Thieme, Stuttgart, 1958, S. Follin *et al.* say: "A differential sign of the highest importance (which we owe to K. Conrad) ... a hysteric is absolutely indifferent to 'anonymous presences', whereas the schizophrenic oniroide attack *(bouffée oniroïde schizophrénique)* attributes thereto, on the contrary, a function immediately significant and projective." See S. Follin, J. Chazaud, and L. Pilon, 'Cas cliniques de psychoses hystériques'; *Evol. Psychiat.* 2 (1961), 257-289; especially p. 275; our translation.

[18] Von Gebsattel is quoted in V. E. Frankl, V. E. von Gebsattel, and J. H. Schultz (eds.), *Handbuch der Neurosenlehre und Psychotherapie,* Urban & Schwarzenberg, Munich and Berlin, 1959; vol. II, p. 311, as saying: "The paranoic lives in a world deprived of harmlessness. Meaningless happenings for him become meaningful. Nothing is mere coincidence, everything is premeditated and so directed at him. Only because of our understanding of the paranoic are we able to realize how fortunate we are that we move in a world of which we are not the main concern, and yet parts thereof; the world moves without paying attention to us, as it were." (Our translation.) See also G. Benedetti, *Der Psychisch Leidende und seine Welt,* Hippokrates Verlag, Stuttgart, 1964, p. 128.

Even in terms of intellectual processes, Piaget's notion of centrism will be another version of this excellent description of the paranoic world. Yet, this very description is so common to all magically minded cultures, as E. E. Evans-Pritchard stresses so much: in magic every event, including what we call accidents, is meaningful, i.e. good or evil.

See J. Agassi, 'On the limits of scientific explanation: Hempel and Evans-Pritchard', *Philosophical Forum* 1 (1968), 171-183.

See also Chapter 3, note 30.

[19] The transformation of the world of the persecuted to the world of the megalomaniac was noted by all the classics, from J. P. Falret to E. Kraepelin; the most detailed is the study of A. Foville *(fils), Mémoire de l'Académie de Médecine*, 1871, pp. 334-350. His famous "I am persecuted, ergo I am a great personality" is quoted in H. Ey, *Études psychiatriques,* Desclée de Brouwer, Paris, 1948-1952; vol. II, p. 533.

There is an added factor to the logic leading from fixation to egomania to persecution to megalomania. It is that in paranoia there is, from the start, an element of egomania, of the readiness to overrule public opinion and decree. This is so regardless of the question, which we consistently keep clear of, of how exactly paranoia occurs and why exactly. Yet, no doubt, the egomania, no less than the intellectual preconditions (having more than one world-view around), are to a large extent cultural determinants. In shamanic society, for example, where the hysterical can become a shaman, there is no room and no need for egomania and so no paranoia. Geza Róheim and Mircia Eliade have already noted both the hysteria of the shaman and the job opening for a hysteric to become a shaman.

[20] Can the dissociation which is subsidiary to the systematic delusion come at once with it? We suggest that this may happen in late adolescence and in senility if and when the fixation takes place despite great fears thereof (understandably) in a violent breakdown with resultant immediate damage (unlike the commoner case of developing a systematic delusion with no accompanying breakdown). The result may be that in the rush and tumult of the breakdown, the patient has a fixation on a silly idea. A youth may think that there is power in words to create things, to resurrect the dead, etc., and so feel utter megalomania at his very psychotic breakdown. Or an old scientist may produce a not quite meaningful formula, which may look almost meaningful, as a new important scientific one often looks at first blush, hope for the best, and in a fearful psychotic breakdown fix a delusion that the best has arrived. We have observed such cases.

[21] An important possibility is opened up by the present discussion. We have noted that paranoia is both rare and the object of most jokes about madmen and stories about madmen and stories about mad scientists. Perhaps this is so because paranoia catches the public's fancy more than other illnesses, regardless of its rarity, perhaps because it is more amenable to fiction, anecdotal or otherwise (indeed, all Ibsen's *Peer Gynt* patients are paranoic, and quite unconvincingly). Yet perhaps this is so because paranoia is commoner than clinically known, simply because between onset and detection, derivative and subsidiary symptoms of dissociation usually develop so fast, that by the time he is seen by the diagnostician the patient is usually wrongly diagnosed as paranoid schizophrenic or as catatonic schizophrenic or as hebephreno-paranoid, etc., whereas his associates, who meet him before his diagnosis, see him mainly

as paranoic proper. The very insistence of diagnosing only pure cases of paranoia as *paranoia vera* already assume that very few will pass the stringent test. It is no accident, then, that though the onset of schizophrenia is spread over all ages beginning in childhood or adolescence, the onset of paranoid schizophrenia is confined to the third decade almost exclusively. We report one case, for example, of late second decade schizophrenia with clear megalomania preceded by an obvious, though not quite pathological case of *idée fixe*. It is quite possible that in reality it was a case of rapidly deteriorating paranoia. This is corroborated perhaps by that patient's variety of symptoms, delusions, hallucinations, catatonia, normal fixations of neurotic character, etc.

[22] It has almost become a custom in psychiatric circles to speak of paranoia in terms of classical psychoanalysis. It is of interest that nonetheless many psychiatrists and psychopathologists see paranoia as a disturbance on a different plain altogether. This, of course, is not conducive to clarity and we recommend a terminological or taxonomic reform.

[23] Homosexuality in Western mixed society is neurotic and all too often distastefully hysterical. There is a persistent effort of certain intellectual homosexuals to present their affliction as the norm, exemplified by animals, or as the super-norm exemplified by Leonardo and Michelangelo. This, however, is never systematic enough a delusion to count as paranoic. Freud's linking of homosexuality and schizophrenia seems highly questionable, since his view of both homosexualism and schizophrenia as narcissistic is questionable. Narcissism, no doubt, can be both normal and a symptom of stress; as a symptom of stress it may accompany all mental ills.

[24] See Chapter 1, note 4.

[25] A. Ehrenzweig says, "Insanity may be creativity gone wrong." See A. Ehrenzweig, 1967, *The Hidden Order of Art: A Study in the Psychology of Artistic Imagination,* Paladin, London, 1970, p. 269. This raises the paradox of paranoia of course. Who is to judge the artist from the madman and by what standards, today's or posterity's? The very dissociation of the madman is no argument; as Ehrenzweig says an artist must disintegrate before he reintegrates creatively in a new mode.

Perhaps even the anxiety of the madman is shared by the artist as a part of his act of creation. There is no doubt that many individuals had a moment of choice: should I now care for my sanity and desist from my work, or should I go on and risk it? And of those who did risk it, some went mad, some not, some created, some not – all four possibilities exist in the history of our culture.

Yet, we think, we have solved the paradox enough to say at what point the creative person went mad: when, due to anxiety, he became unable to cope with all the parameters he had to and used to operate with. Mental illness is the illness of the thinking apparatus, at least in psychosis if not also in severe neurosis.

See also Chapter 7, note 3 and note 26, below.

[26] The paranoic has a fixation on an abstract system. He has the same number of parameters simultaneously present to his mind, as the young adolescent. If one has a fixation on an abstract system, but a higher number of parameters present simultaneously in one's mind, one is not paranoic but fanatic, etc. A mere lower number of parameters, (provided that there is no fixation) on the

other hand, may make one a primitive, or a retarded person, etc.

[27] R. Holt says in D. Rapaport, M. Gill, and R. Schafer, *Diagnostic Psychological Testing* (ed. by R. Holt), University of London Press, London, revised ed., 1968: "What, then, is the role of diagnosis (in the sense of using a nosology) in clinical practice? A diagnosis is not a sufficient classification but a necessary constituent of a personality description. In order to explain how it is used, it will be necessary to state a conception of how typological concepts such as diagnoses are useful in a world of continuous variation..." (p. 14).

"... The diagnostician will have to stick to the landmark approach, using the classical diagnostic categories as orienting points. Occasionally, a person may be close enough to the center of the hypothetical region defining a diagnostic entity so that he can be located almost entirely by reference to it, without discussing the 'next of kin'; in this limiting case only, diagnosis can approach pigeonholing. Usually, however, the diagnostician will have to triangulate by reference to several such landmarks" (p. 14-15).

This is why we did not here present any case histories, since the 'typical case' is admittedly an abstraction.

Consider, for example, the case, reported a few times, of a person who suffers from paranoia whose delusion he has acquired in a single incident of hallucination; he has not hallucinated since. Is his case *paranoia vera* or schizophrenia? Suppose it is the latter. Is it still so if that hallucination was chemically induced?

See also note 22 to Chapter 2 above.

[28] The difficulty with the loose use of the term persecution in English was noticed by J. Hoenig and M. W. Hamilton, the English translators of Jaspers, in their preface to K. Jaspers, 1913-1945, *General Psychopathology,* University of Chicago Press, Chicago, 1963, p. viii: "The term 'paranoid' presented us with special difficulties. We have used it in keeping with the Oxford Dictionary definition of Paranoia (n. Mental derangement, esp. when marked by delusions, of grandeur, ...) The German usage is also in line with this definition and implies a 'wrong notion', synonymous with 'delusional'. In English, however, 'paranoid' is often used to mean 'persecutory' both in technical psychiatry and in general usage. This has led to a good deal of confusion with such expressions as 'paranoid schizophrenia' (i.e. a schizophrenic picture in which delusions predominate), since the content of the delusions can have other than persecutory content, e.g. messianic, hypochondriacal, magical, etc. Thus the expression 'paranoid features' does not properly mean 'ideas of persecution' but merely 'the presence of delusions'. As there does not seem any philological justification for the use of this word in the narrower sense of 'persecutory', we have retained the broader meaning in accordance with the dictionary definition."

The confusion of psychoses of diverse sorts with persecution, even the identifying of both, is very common, in folk mythology and in the clinical literature alike. Consequently much has been written on persecution, and we are now extremely familiar with its normal manifestations, make believe manifestations, delusional manifestations, and even cultural ones. Indeed, this last one, the persecutionist ethos, is now increasingly fashionable, and we comment on it in note 30 to Chapter 3 above.

What is missing in all this is the neurotic persecution, as opposed to the psychotic one, even though its mechanism is classical: the projection of anxiety to the environment. The sole observer of it as neurotic, but non-psychotic, is Bertrand Russell, who has a chapter on it in his celebrated *The Conquest of Happiness,* where he avoids discussing extreme cases yet notices that mild cases are common enough. They are rooted, he says, in an exaggerated self-importance.

The important theoretical point we make is that the mechanism of psychotic persecution is different from that of the neurotic one: it is the effort to stick to an *idée fixe* at all cost, especially to the delusion that it is public knowledge (e.g. not merely an illusion of grandeur, but of obvious grandeur, and requiring an explanation of the absence of homage – and obviously persecution is an easy mode of explanation). The reason neurotic persecutionism is ignored, is that many authors, e.g. Melanie Klein, usually ascribe as neurotic mechanism to psychosis. Our main concern is precisely to distinguish between the two mechanisms, although we also wish to add, incidentally, that the mild persecution mania is more often neurotic than psychotic, and that psychotic persecution mania is a corollary to paranoia, mild or severe. Also, we would add, neurotic persecution mania is usually a projection of an ambivalence concerning one's loneliness. Also we should add, all persecution mania relates to the suffering from loneliness and the inability to face it squarely and either tolerate it or do something about it.

Finally, consider a condition which is common in normal people no less than in patients, namely hypochondria, of which Kant says in his *Anthropology* that it is between sanity and insanity (so that there is no clear-cut division – *deutlicher Abschnitt* – between them).

Hypochondria is usually classed as neurotic on account of its mildness; which is not a good reason, of course. We agree that it is often neurotic in that it is so evidently symbolic; it is decidedly also often psychotic, in being an *idée fixe* about one's own self, which is often persecutionist and, as Kant notices, avidly reinforced by reading medical texts and finding – this is more common than hypochondria, even – that one has all the symptoms and illnesses one reads about. Hypochondriac phantasies are both intellectual and emotional, and so, perhaps hypochondria is on the borderline between, or a combination of, psychosis and neurosis. The reason that even non-hypochondriacs show hypochondriac signs when reading a medical text has already been given here: in distress one relates everything to one's own condition, and laymen usually consult a medical text under mounting pressure. Again we say, paranoia is a healthy sign of anxiety run wild; and so hypochondria which is almost normal may become *paranoia vera.*

APPENDIX I

1 E. Bleuler, 1911, *Dementia Praecox or the Group of Schizophrenias,* International Universities Press, New York, 1950: "Let us begin by saying that the secondary symptoms [of schizophrenia] are a direct consequence of the loosening of the associations" (p. 352).

2 *ibid.,* p. 298: "Definite schizophrenic disturbances of association alone, are sufficient for the diagnosis [of schizophrenia]".

3 O. Fenichel, 1946, *The Psychoanalytic Theory of Neurosis,* Routledge & Kegan Paul, London, 1966: "There is a definite order in [schizophrenic] thinking . . . schizophrenic logic is identical with primitive, magical thinking . . ." (p. 421) " . . . It is the archaic way of thinking" (p. 421).

[4] See H. Werner, 1948, *Comparative Psychology of Mental Development,* Science Editions, New York, 1961. Werner's monograph refers to animal behaviour, ethnology, child psychology, and psychopathology, as well as to works by H. Kluver on mescalin and by K. Beringer both on schizophrenic thinking and on mescalin.

[5] L. Lévy-Bruhl, *Les fonctions mentales dans les sociétés inférieurs,* Alcan, Paris, 1910.

[6] J. Piaget, 'La pensée symbolique et la pensée de l'enfant', *Archiv de Psychol. de Genève* 18 (1923), 273-304. It is to be noted that though Piaget here describes symbolic thinking and child's thinking as similar, he never confuses them nor sees them as identical.

[7] S. Freud, 1912-1913, *Totem and Taboo: Some Points of Agreement between the Mental Lives of Savages and Neurotics.* Hogarth Press, London, *S.E.* 13 (1955), 1-161.

The idea that neuroses are archaisms is implicit in Freud, but is explicitly ascribed to him by Geza Róheim. See the discussion on Róheim in J. Agassi, *Towards a Rational Philosophical Anthropology,* 1976.

Clearly the idea includes the seductive but suspect theory that ontogeny recapitulates phylogeny, together with Freud's view of dreams and day dreams as neurotic, as well as Freud's view of psychosis as severe neurosis – a view he held all his life; see Appendix II. Yet in his posthumous *An Outline of Psychoanalysis,* [1938], 1940, Hogarth Press, London, *S.E.* 23 (1975), 141-207, part II, Chapter VI, first para., when discussing dreams, he deviates with no prior warning. It is obvious that for him dreams and day dreams are neurotic (see note 27 to Chapter 9 below); that Jung's view of them as psychotic was deviant, and that Bleuler, following Jung here (see next note), was part and parcel of his leaving the orthodoxy. Yet here Freud speaks in the Jung-Bleuler vein:

"A dream, then, is psychosis, with all the absurdities, delusions and illusions of a psychosis. A psychosis of short duration, no doubt, harmless, even entrusted with a useful function, introduced with the subject's consent and terminated by an act of his will. None the less it is a psychosis, and we learn from it that even so deep-going an alteration of mental life as this can be undone and can give place to the normal function. Is it too bold, then, to hope that it must also be possible to submit the dreaded spontaneous illnesses of mental life to our influence and bring about their cure?

We already know a number of things preliminary to such an undertaking. According to our hypothesis it is the ego's task to meet the demands raised by its three dependent relations – to reality, to the id and to the super-ego – and nevertheless at the same time to preserve its own organization and maintain its own autonomy. The necessary precondition of the pathological states under discussion can only be a relative or absolute weakening of the ego which makes the fulfilment of its tasks impossible. The severest demand on the ego is probably the keeping down of the instinctual claims of the id, to accomplish which it is obliged to maintain large expenditures of energy on anticathexes. But the demands made by the super-ego too may become so powerful and so relentless that the ego may be paralysed, as it were, in the face of its other tasks".

For a detailed analysis of all this see Appendix II. Now, clearly, Freud reverts here, not so much to dreams, nor to whether they are pathological etc., but rather to his old vexing problem, can psychoanalysis help cure psychosis. See also note 23 to Chapter 2 above about this.

We can scarcely avoid noticing, perhaps with some unjust glee, that when Freud views a day dream as paranoic in spite of the day dreamer's full control, he is hit by the paradoxes of paranoia beyond salvation. Yet, we add, shamefacedly, taking Freudianism as one integrative principle barely permits deviation from this mere corollary to his fascinating, grand equation of the archaic, primitive, infantile, illusory, dreaming, neurotic, psychotic, and even creative. For the allure of this grand equation see the last paragraph of the next note.

[8] See E. Bleuler, 1911, *Dementia Praecox or the Group of Schizophrenias*, International Universities Press, New York, 1950. E. Bleuler dealt with the problem of "the Relation of Schizophrenia to Dreams" in pp. 439-411 of his famous monograph. He approaches the problem in three different ways: one, by comparing schizophrenics' delusions with schizophrenics' dreams; two, by comparing schizophrenics' delusions with normal peoples' dreams; three, by comparing schizophrenics' dreams to healthy persons' dreams. His conclusions are these. As for the first comparison, i.e., schizophrenics' autism and schizophrenics' dreams, it is that "thinking in schizophrenics' dreams and schizophrenic autistic thinking, are essentially identical" (p. 440). Here Bleuler adds that his findings "are confirmed by Kahlbaum, Kraepelin, Sante de Sanctis, and von Krafft-Ebing" (p. 440 note). As for the second comparison, i.e., between schizophrenics' delusions and healthy persons' dreams, Bleuler's conclusion is that "in spite of the difference in genesis and in spite of other minor differences, it may yet be possible to show that the secondary symptomatology of schizophrenia [i.e. delusions] is wholly identical with that of dreams". As for the third comparison (i.e., between schizophrenics' dreams and healthy persons' dreams), Bleuler says: "It should also be mentioned here that those dreams of schizophrenics which have been analysed up to now differ in no way from the dreams of healthy persons" (p. 440).

The problem we are going to raise here, namely, "if dreaming is the same as delusions, then why have both?" is not touched upon by Bleuler.

Additionally, referring to schizophrenic phenomena and the dream, Bleuler says, *op. cit.*, p. 440, that "the only difference I could see until now between schizophrenic phenomena and the dream, is in the most pronounced dislocation *(Spaltung)* of the personality. The dreamer is dominated by a homogeneous mixture of complexes. The schizophrenic processes by a double registration . . . in the sense of reality and in the sense of the delusion. The difference is however not essential."

Perhaps the difference lies in the levels of organization of the event, momentary or permanent as it may be; this difference is essential for the differentiating between dreams and delusions, as well as from the whole range of the other conditions mentioned in the beginning of Appendix I.

Yet we may add here that the secret of the allure of the equation of all mental aberrations (see previous note), from day dreams to stark madness, is surely in significant true instances to it. In this vein, E. H. Gombrich conjectures that plans often mature out of day dreams, and even originate there. He even thinks language stems from day dreams (the yam-yam theory, so called). See his 1963, *Meditations on a Hobby-Horse and Other Essays on the Theory of Art*, Phaidon Press, London, 1965.

[9] For the child's conception of dreams, see J. Piaget, 1926, *The Child's Conception of the World,* Harcourt, Brace, New York, 1929.

[10] A. Wheelis, *The Quest for Identity,* Norton, New York, 1958.

[11] See J. Breuer and S. Freud, 1893-1895, *Studies on Hysteria,* Hogarth Press, London, *S.E.* 2 (1955), repr. 1957, 1962.

[12] See K. Goldstein, 1940, *Human Nature in the Light of Psychopathology,* Schocken, New York, 1963, where his Chapter 2 begins with explicit reference to the importance of the work by A. Storch, 1922, 'The Primitive and Archaic Forms of Inner Experiences and Thought in Schizophrenia', *Nerv. and Mental Disease Monog.* 36 (1924), and following it, examples of brain injured patients studied by Goldstein, are reported.

See also K. Goldstein, *Language and Language Disturbances,* Grune & Stratton, New York, 1948, where the book begins with references to the importance of Piaget's works to the study at hand.

[13] A warning against reducing psychiatry to neurology is found in both Freud and Jackson – the two pioneers of the modern concept of aphasia. See S. Freud, 1891, *On Aphasia: A Critical Study,* International Universities Press, New York, 1953. In Chapter V, p. 56 Freud notes: "Hughlings Jackson has most emphatically warned against such a confusion of the physical with the psychic in the study of speech [and he quotes Jackson, thus]: 'In all our studies of diseases of the nervous system we must be on our guard against the fallacy that what are physical states in lower centres fine away into psychical states in higher centres; that, for example, vibrations of sensory nerves become sensations, or that somehow or another an idea produces a movement'. *Brain* 1, p. 306."

[14] For a recent excellent overview and critical study of the problem of impaired thinking in schizophrenia covering all its aspects (logic, cognition, perception, emotion, etc.), see L. J. Chapman and J. P. Chapman, *Disordered Thought in Schizophrenia,* Prentice-Hall, Englewood Cliffs, N. J., 1973, as well as for personal contribution of the same authors to the problem at hand.

Some of the more interesting works:

Logic: E. von Domarus, 'The Specific Laws of Logic in Schizophrenia', in J. S. Kasanin (ed.), *Language and Thought in Schizophrenia: Collected Papers.* University of California Press, Berkeley, 1944, as well as other papers in this collection; S. Arieti, 1955, *Interpretation of Schizophrenia,* Brunner, New York, enlarged ed., 1974; A. Matte-Blanco, 'A Study of Schizophrenic Thinking: its Expression in Terms of Symbolic Logic', *Congress Report,* Vol 1, *International Congress of Psychiatry, Zurich,* 1957.

Conceptual Thinking: See notes 15, 16, and 17 below.

Overinclusion: N. Cameron, *Personality Development and Psychopathology,* H. Mifflin, Boston, 1963; R. W. Payne, 'The Measurement and Significance of Overinclusive Thinking in Retardation and Schizophrenic Patients', in P. Hock and J. Zubin (eds.), *Psychopathology of Schizophrenia,* Grune & Stratton, New York, 1966.

Cognition: G. Miller, 1956, 'The Magical Number Seven', Chapter 2 of his *The Psychology of Communication,* Basic Books, New York, 1967; (this classic essay is a bit of a shaggy dog, yet a must); D. Shakow, 'Segmental Set: A Theory of Formal Psychological Deficit in Schizophrenia', *Archs. Gen. Psychiat.* 6 (1962), 1-17; A. McGhie, *Pathology of Attention,* Penguin, Harmondsworth, 1969.

Perception: G. S. Klein, 'The Personal World Through Perception', in Blake and Ramsey (eds.), *Perception: An Approach to Personality,* Ronald Press, New York, 1951. Klein states: "Perception is *the* point of reality contact, the door to reality appraisal, and there is no doubt that here especially are the selective controls of personality brought into play" (p. 328). (See also G. S. Klein, *Perception, Motives and Personality,* Knopf, New York, 1970.)

While G. Klein speaks of personality, J. Piaget refers to perception as part of regulative intelligence-like mechanism. J. Piaget, *Les mécanismes perceptifs,* Presses Universitaires de France, 1961. Interesting links can be drawn here between personality, intelligence, perception and disease.

A clear and general picture of schizophrenic thinking is drawn by B. Maher, 'The Language of Schizophrenia: A Review and Interpretation', *Brit. J. Psychiat.* 120 (1972), 3-17.

[15] L. S. Vigotsky, 'Thought in Schizophrenia', *Archiv. of Neurol. and Psychiat.* 31 (1934), 1063-1077, L. S. Vigotsky, *Thought and Language,* Wiley, London and New York, 1962.

[16] See E. Hanfmann and J. Kasanin, 'Conceptual Thinking in Schizophrenia', *Nerv. and Mental Disease Monogr.* 67, 1942.

[17] K. Goldstein and M. Scheerer, 'Abstract and Concrete Behaviour: An Experimental Study with Special Tests', *Psychological Monographs* 53 (1941), 1-151.

[18] N. Geschwind, 'Anatomy and Higher Functions of the Brain', in R. S. Cohen and M. W. Wartofsky (eds.) *Boston Studies in the Philosophy of Science,* Vol. 4, Reidel and Humanities, Dordrecht and New York, 1969, pp. 98-136. See also his *Selected Papers on Language and the Brain,* same series, vol. 16, 1975.

[19] See note 14, above.

[20] See Chapter 9, note 22.

[21] See P. Pichot, *Les tests mentaux en psychiatrie,* Presses Universitaires de France, Paris, 1949; D. Rapaport, M. Gill, and R. Schafer, *Diagnostic Psychological Testing,* R. Holt (ed.), University of London Press, London, rev. ed., 1968.

[22] K. R. Popper, *The Poverty of Historicism,* Routledge & Kegan Paul, London, 1957.

[23] P. B. Medawar, 1959, *The Future of Man,* Mentor Books, New York, 1961.

APPENDIX II

[1] S. Freud, 1938, 1940, *An Outline of Psychoanalysis,* Hogarth Press, London *S.E.* 23 (1964), 140-207.

[2] S. Freud, 1924, *Neurosis and Psychosis,* Hogarth Press, London, *S.E.* 19 (1961), 149-153.

[3] S. Freud, 1924, *The Loss of Reality in Neurosis and Psychosis,* Hogarth Press, London, *S.E.* 19 (1961), 183-187.

[4] S. Freud, 1927, *Fetishism,* Hogarth Press, London, *S.E.* 21 (1961), 152-157.

[5] See note 2, above.

[6] This is also stated in Freud's Famous *Dictionary,* Art. 'Paranoia'.

[7] See note 3, above.

[8] See note 2, above.

[9] See note 4, above.

[10] See notes 2 and 3, above.
[11] A. Koestler, *The Sleepwalkers,* Grosset & Dunlop, New York, 1959.
[12] See note 1, above.
[13] See E. Jones, *Sigmund Freud: Life and Work,* Hogarth Press, London, vol. 3, 1955, p. 255; Basic Books, New York, vol. 3, 1955, p. 239.
[14] S. Freud, 1938, 1940, *Splitting of the Ego in the Process of Defense,* Hogarth Press, London, *S.E.* 23 (1964), 275-278.

CHAPTER 9

[1] For demarcation, see D. Bohm's notion of order: D. Bohm, 'Some Remarks on the Notion of Order', in C. H. Waddington (ed.), *Towards a Theoretical Biology,* Edinburgh University Press, Edinburgh, 1968; Spencer Brown's essay in formalization: G. Spencer Brown, *Laws of Form,* Allen and Unwin, London, 1969, 2nd impr., 1971; and Golani's methodological study of qualitative phenomena: I. Golani, 'A Choreography of Display', Ch. 2 in P. Bateson and P. Klopfer (eds.), *Perspectives in Ethology,* Vol. 2, Plenum, New York, 1976.

In psychiatry: for Lacan's topological approach, see J. Lacan, 'Paranthèse des paranthèses', in his *Ecrits,* éd. du Seuil, Paris, 1966. For works of Pichot's school, in applying a quantitative approach to demarcation, see J. E. Overall, L. E. Hollister, and P. Pichot, 'Major Psychiatric Disorders: a Four-Dimensional Model', *Arch. Gener. Psychiat.* 16 (1967), 146-151.

[2] See T. S. Kuhn, *The Structure of Scientific Revolutions,* Chicago University Press, Chicago and London, 1962, second ed., 1970; reviewed by J. Agassi in *J. Hist. Philos.,* 4 (1966), 351-354. See the standard discussion of this topic, especially the contribution of Margaret Masterman, in I. Lakatos and A. Musgrave (eds.), *Criticism and the Growth of Knowledge,* Cambridge University Press, Cambridge, 1970, reviewed by J. Agassi, 'Tristram Shandy, Pierre Menard, and All That: Comments on *Criticism and the Growth of Knowledge*', *Inquiry* 14 (1971), 152-164. See also J. Agassi 'Sociologism in the Philosophy of Science', *Metaphilosophy* 3 (1972), 103-122.

Kuhn speaks of a pre-paradigm discourse as pre-scientific. Taking seriously Claude Lévi-Strauss' view of primitive thinking as the science of the concrete, as purely associational and otherwise non-integrative, we can conclude that primitive people cannot be psychotic – even if they have delusions. This is certainly not true, as we learn from E. E. Evans-Pritchard, and as we have repeatedly noted in previous chapters.

If primitive people have paradigms, will Kuhn call them scientists? He will have to, as he admits even mediaeval theologians are. This is a problem with his demarcation of science, however; not with our demarcation of psychosis. We insist that our demarcation reopens the disputed question, can primitives be psychotic, in a new way, and may enable us to reach agreement on it by empirical means.

[3] I. Lakatos, 1970, 'Falsification and the Methodology of Scientific Research Programmes', in I. Lakatos and A. Musgrave (eds.) *Criticism and the Growth of Knowledge,* Cambridge University Press, Cambridge, 1970, pp. 91-196. Also: I. Lakatos, 'History of Science and its Reconstructions' in R. C. Buck and R. S. Cohen (eds.), *Boston Studies in the Philosophy of Science,* vol. 8, Reidel,

Dordrecht and Boston, 1972, pp. 91-136.
[4] We employ here a generalized Jacksonian principle, because for Jackson the secondary signs are of releases of already existing lower integrative principles, whereas we allow ourselves to speak of such lower principles as also evolving to compensate for the primary defect, rather than merely being released. This is in accord with both Jackson and Freud. See Chapter 1, note 21, notes 16 and 23.
[5] Of course, dissociation may occur as a primary organizational defect. We call it dementia and mean to declare it caused by brain-damage. Whereas the brain-damage theory (no neurosis without psychosis) is originally metaphysical, our distinction between dementia and mental illness amounts to its refutation, and so to the view of it as empirical. All this is in very strong agreement with Jackson's principles.
[6] S. Arieti, in his *Interpretation of Schizophrenia,* Brunner, New York, 1955, suggests *the principle of teleological regression* in order both to describe and understand behaviour: "*regression,* because less advanced levels of mental integration are used; *teleologic,* because this regression seems to have a purpose, namely, to avoid anxiety by bringing about the wanted results" (italics in the original, p. 192). We fully endorse this; see note 8 below.
[7] That all mental patients suffer from anxiety does not imply the converse, i.e. that all those who experience anxiety are mental patients. Moreover, we know of at least one form of (experiencing) anxiety, where it is a sign of mental maturity and health rather than a sign of illness. See J. J. Lopez Ibor, 'Angoisse, existence et vitalité', *Evolut. Psychiat.,* 263 ff., 1950. See also Y. Fried and F. Brüll, 'Intensive Psychotherapy for Acute Psychiatric Patients: Theoretical Considerations', *Brit. J. Psychiat.* 121 (565); 635-639 (1972).

We do not consider all psychopaths to be mental patients in the medical sense; our opinion is that they might be socially labelled 'ill', or even 'sociopaths,' but not medically so. It is precisely because of the lack of anxiety in some cases of psychopathy that we deny it the status of a psychiatric condition proper.
[8] Fixation is a projection of a problematic emotion, we say. We may call it also an emotional problem, except that emotional problems when contrasted with intellectual problems tend to be ambiguous in the following way. An intellectual problem may be utterly impersonal, as Fermat's problem in mathematics is, or as any decision problem in logic or in economics. An intellectual problem as opposed to an emotional problem may be a problem regarding the patient's intellect as opposed to a problem regarding the patient's emotions. Now the intellectual problems in the abstract sense are at times remote from the patient and he ignores them or attempts to solve them in the course of his study or research. Or they may be problems threatening his delusion, in which case he offers a facile and quick solution, just as the fanatic does (see note 11 to Chapter 7). But the problem with his intellect, in this case, is secondary to the problem with his emotion which he has tried to solve by damaging his intellect. See note 6 above.
[9] The importance of symbols as abstract entities, yet such that have some connection – onomatopoeic or associational – with their designata, seems to be what H. Werner and B. Kaplan, try to bring to our attention in their scholarly *Symbol Formation, an Organismic-Developmental Approach to*

Language and the Expression of Thought, Wiley, New York and London, 1963. But we cannot be certain since the volume is enormously eclectic and we are uneasy about it. For example, the authors accept ideas from Kraepelin and Federn and Melanie Klein, perhaps without involving themselves in inconsistency, perhaps not; and they cite O. Jespersen whose views are undoubtedly different from theirs, and W. Sterne and R. Jacobson and others, as if everyone agrees with them. However, they do observe a few important items in the respect mentioned, and we should note them.

The authors note that the earliest stage of language acquisition is onomatopoeic and the next is of "progressive distancing" (p. 205), which may finally perhaps destroy all "inner bond between vehicle and referent." i.e. the link between a symbol and its meaning may finally become very loose. (Meanings of names and of descriptive phrases, according to Frege and Russell, and almost all logicians agree in broad outline, is sense plus reference. We are not clear what theory of meaning the authors hold, except that meaning involves reference.)

The authors note that there are degrees of distance between symbol and its referent, especially since the color of the symbol's meaning can be both taken away from it and injected back into it (especially in dreams, in associating word and gesture, etc.).

The lack of all distance between word and thing (as if all early acquired words name things; cf. pp. 137-9 there; but never mind that) is the treatment of a word "as if it were a thing" (p. 250), or "word-realism" – particularly in dreams. The authors refer to Freud's (1900) *Interpretation of Dreams* here, as well as to Kraepelin, *Uber Sprachstörungen im Traum* of 1906. Viewing this "word-realism" as a regression to an early stage, and viewing schizophrenia as essentially autistic (p. 253) (the authors refer here to E. Bleuler's (1911) *Dementia Praecox or the Group of Schizophrenias*), they treat the two on a par.

We are not clear ourselves why autism makes this connexion between dreams and schizophrenia. To make things harder on us, the authors next endorse Federn's (1952) theory of schizophrenia as the loss of sense of body-boundary, because they consider it regressive too, saying (in accord with Federn, to be sure, though it is an aside for him) that it is "reminiscent of the condition obtaining in early infancy and in dream states" (p. 254). This is really far fetched because bed-wetting answers this description no less, and perhaps more, than the alleged loss of bodily boundary feeling; moreover, the view that this condition is common in infancy is a theory (of Melanie Klein), whereas bed-wetting is an attested fact.

Be it as it may, the authors now refer to Hans Kreitler's report ('Les bases psychologiques du language des schizophrènes' *Acta Neurol. et Psychiat. Belgica* 57 (1957), 950-954) that schizophrenics always involve their auditors in their fantasies ("always" is much too strong here), and, they add, schizophrenics are generally confused and so they tend also to shrink distance between words and things.

We hope that our disagreement with these authors is obvious enough. We speak not of abstract symbols and their referents but of unifying abstract principles, something not of early infancy but of late adolescence, not something which is part of the general psychotic confusion but a fixation which is at the center of it.

[10] The psychology of religious experience loosely known as the mystic union has been analysed by Freud as the inversion of insecurity into impotence turned into omnipotence. He does not say what differentiates mystic omnipotence from technological omnipotence, and he identifies paranoid schizophrenic delusions as similar to mysticism without saying who develops into mystic, who into schizophrenic. See J. Agassi, 'Unity and Diversity in Science', in R. S. Cohen and M. W. Wartofsky (eds.), *Boston Studies in the Philosophy of Science,* vol. 4, Reidel, Dordrecht and Humanities, New York, 1969, pp. 463-522, especially p. 467, reprinted in his *Science in Flux, Boston Studies,* vol. 28, 1975.

Now clearly, psychologically speaking, the mystic projects his "positive" self onto the whole world, whereas the psychotic's projection is more limited. Moreover, we suggest, catatonia is achieved by projecting one's anxiety as an integrative principle and when it grows far enough (to become a symbol and also quite embracing) introjecting it so that the world becomes one with the subject, not as in the mystic experience where the subject becomes one with the world.

Nevertheless, we suggest, though the patient internalizes the world, and though the affective element externalized and reinternalized is anxiety, still there is the joy in catatonia, both the joy of sensing anxiety in its highest purity and to its fullest, and the joy of security, of piercing uncertainty and thereby arriving at certitude (à la Descartes, no less); see R. H. Popkin (1960), *The History of Skepticism from Erasmus to Descartes,* Utrecht and Humanities, New York, 1964; and remember both R. D. Laing's comment on the affinity between quaint philosophy and psychotic delusion, as well as Kierkegaard's expression in his *Fear and Trembling,* of admiration not of Descartes' philosophy but of his courage to undergo the experience of total doubt, of controlling the whole universe and of divining that it is nothing but fear.

Also, the psychotic sense of omnipotence and plethora (not to be confused with his megalomania which is persecutionist) is often an inversion of catatonia, in the Freudian sense mentioned at the opening of this note. Contrast this with Mortimer Ostow, 'The Basic Process of Schizophrenia'. *Dis. of the Nervous System,* suppl., 29 (1968), 16-21, especially p. 20.

Melanie Klein views all projection as that of a previously introjected "bad breast". She views this as paranoid, meaning persecutionist, because "bad breast" is a cause of anxiety.

We hold a different view and think that any intellectual fixation is paranoid and is caused by anxiety and the cause of both anxiety and persecution. Yet we find no inconsistency between these two views, and in accord with Klein may suggest considering catatonia as an introjection of the "bad breast" for the second time around, after it has been projected and systematized to cover the whole world. Whether this offers an insight into the passion to make the systematic principle universal and thus truly integrative as a regression to the stage when the breast was the whole of the outside world, this depends on one's viewpoint. We prefer to consider this passion as an expression of one's faith in one's superior logic and a justification for one's disregard of public opinion. But, again, there is no inconsistency here between our own view and Klein's.

[11] When discussing a quantitative gradation and a qualitative demarcation, the

general question of discontinuity *versus* continuity is immediately raised, and rightly so. Yet we think it is premature to extend what we have to say in this context about the general and obviously metaphysical problem.

[12] It is generally agreed that abstract thinking is higher than concrete thinking. Now, we could claim that neurosis is structurally lower than psychosis. But the concrete fixation in neurosis is local, and therefore there is no place to consider neurosis as a lower organizational level of functioning than psychosis. Psychosis, is an abstract fixation (global) of a given rather low level, whereas neurosis is symbolic fixation (local) within the highest level of integration of the patient. See Chapter 1, note 16.

[13] See Chapter 7, note 23.

[14] See Chapter 3, note 25.

[15] In the conclusion to his *To Define True Madness,* Henry Yellowlees, Penguin [Harmondsworth, 1946] defines madness as an escape to a private paradise – and we agree – where entry is free for children and for young people at half-price. Not bad at all.

[16] A detailed study of the case of a borderline psychotic with the signs and symptoms mentioned is to be found in A. J. A. Symons, *The Quest for Corvo,* Michigan State University Press, East Lansing, 1955. It is a detailed, vivid, and kaleidoscopic picture of Frederick Rolfe, self-styled Baron Corvo, author of *Hadrian the Seventh* (1904), recently successfully made into a play. Corvo is clearly presented as a borderline psychotic (pp. 93, 99) with fixed ideas (p. 212), with delusions of grandeur and persecution (pp. 106, 240), with strong narcissism (p. 62) as well as with utter horror of sex and of any other form of intimacy (pp. 68, 88, 150-1, 153, 237-8), with pathological clinging to hosts while forcing them to throw him out, with restless wandering, etc. He could even fall into short spells of catatonia. Yet he failed to be certified (p. 42) and today it is even less likely that he could be found certifiable. Clearly most of his illusions about original inventions were accompanied by some slight familiarity with the field within which he claimed to have made a contribution, so he was oft on the borderline between the psychotic and the charlatan (as even many talented, attested charlatans are), if not a charlatan proper (pp. 89, 185, 211).

Solomon Maimon describes in his autobiography what we would consider a rather prolonged psychotic episode, which he describes as his period of wandering as a beggar. He confessed he could not explain what made him become a beggar, except, partly, his failure to be accepted into the foreign community which he had tried to join. See Solomon Maimon, 1792, *Salomon Maimons Lebensgeschichte,* Chapter 22.

[17] That depression may be neurotic and psychotic is a well-known fact. We think we can offer a very simple diagnostic test differentiating the two. As we have said above (§ 6), the neurotic has little or no insight into his condition though he is very insightful when seeing it in others; the psychotic tends to be the other way around.

This is in no contrast to the clinical diagnostician's observation – or criterion – according to which a neurotic who suffers from a phobia is aware of it but not a psychotic. That is to say, a neurotic in depression, or any other distress, is aware of his distress but fails to connect it with its background even if he will make the connection in other patients' cases; the psychotic will see nothing wrong with himself, and explain his condition intelligently by reference to his own background but will not be able to perceive others who suffer the same condition.

What is common to both types of depression is expressed in the rather commonplace idea that all depression, neurotic or psychotic, severe or mild, is a simple defense mechanism whose function is to prevent the patient from causing severe damage to himself. A vivid autobiographic picture thereof is given by A. Wheelis, in his *The Quest for Identity,* Norton, New York, 1958. It has been an object for detailed and lengthy study by Mortimer Ostow, in his *The Psychology of Melancholy,* Harpers, New York, 1970. On p. 93ff. he speaks of psychotic depression as an entity apart (and subsidiary).

[18] See note 15, above.

[19] In the summary of his *Melancholie und Manie,* Neske, Stuttgart, 1960, pp. 135-140. L. Binswanger wonders if there really is a difference between delusion *(wahn)* as it occurs in schizophrenic and in manic-depressive psychosis.

See next two notes.

[20] There is the possibility that the integrative principle of manic-depression is not common to both states, but peculiar to one state, where the other state is a mere Freudian inversion of it. Logically, if this is true and if the case exhibits some periodicity, the periodicity will need another integrative principle, perhaps in the form of an *ad hoc* amendment to the initial one. We do not know of such cases but they are possible.

[21] See L. Binswanger (1931-1932), 'Über Ideenflucht', *Archives Suisses de Neurologie et Psychiatrie,* vols. 28, 29 and 30, reprinted L. Binswanger, *Über Ideenflucht,* Miehans, Zurich, 1933. Quoted and passages translated in H. Ey (1948-1954), *Etudes psychiatriques,* Desclée de Brouwer, Paris, vol. III, 1954; pp. 70-87; also L. Binswanger, *Melancholie und Manie,* Neske, Stuttgart, 1960.

The inexorable logic of the manic depressive lies really in both his strong emotions and his evasion of them: the stronger the emotion the stronger the logic, the paranoic thinking that diverts his own attention from it. Yet the thought, to beat emotion, must be very emotionally charged, and so devastatingly increasingly self-defeating, exciting the patient into mania or slowing him to stupor. The explosive nature of the process near the poles is a well observed fact.

[22] Autistic thinking is classically considered to be a (kind of) concrete thinking. We wonder if it is so. When we examine the so-called 'concrete thinking' we immediately notice that there is a whole hierarchy of levels of concrete thinking; from the concrete, literal meaning of a proverb, to the very concrete meaning of a very concrete object, to the concreteness of the aphatic's language. In other words, the hierarchy of the concrete languages is the expression of a hierarchy of diminishing levels of abstraction. Thus, when speaking of autistic thinking as a 'private language', we agree; when speaking of autistic thinking as a 'concrete language' one has to be more cautious and either specify the degree of abstraction which this concrete language consists of, or – better – to speak of degrees of concreteness, complementary to degrees of abstraction.

[23] See F. Brüll, 'The Trauma: Theoretical Considerations', *Israel Annals of Psychiatry* 7 (1969), 69-108.

[24] S. Freud, 1910, *Five Lectures on Psychoanalysis,* Hogarth Press, London,

S.E. 11 (1962), 9-56, p. 46: "an excessively strong manifestation of these instincts at a very early age leads to a kind of partial *fixation* [italic in the text], which then constitutes a weak point in the structure of the sexual function. If in maturity the performance of the normal sexual function comes up against obstacles, the regression that took place during the course of development will be broken at the precise points at which the infantile fixations occured." It is clear that if there is no regression, a mental disease would not take place. Note Freud's remark, "if in maturity," etc. – in other words: in adolescence or post adolescence.

[25] There is, surprisingly, no Piagetian version of the Freudian Oedipus event. This can be worked out easily. For, at the Freudian Oedipal stage the Piagetian infant emerges from the state where he can only operate with one viewpoint, the egocentric one, and so relates things to himself alone and has no sense of reciprocity, let alone between two who are both not himself. And so his attitudes to parents are separate, he can be at times well disposed towards a parent, at times not, but either way with little difficulty (the difficulty and ambivalence come later). When he starts to note the two parents at once, he can have an Oedipal experience; he has also at that time to note a reciprocity between parents. Of course, both Piaget and Melanie Klein stress the importance of the day when the infant recognizes his mother as a person. Yet even on this point information is not compared or collated.

[26] We consider split personality as such, just as the ability to use two integrative systems, quite favourably. Psychiatrists who declare that a strong identity is a necessity for mental health are only right to the extent that split-personality may get out of control and so leading to disintegration, dissociation, etc. For obviously, the patient who operates on merely one integrative principle is better integrated; but he may still be unwell; whereas a split person who is in control may be well enough.

[27] Without any attempt to belittle Freud's theory of dreams and of day dreams we may note that all the dreams he analyzes and all the interpretations of dreams he analyzes as psychopathological (some are perfectly normal) go in the neurotic direction of concrete symbolism. That some dreams can have more intellectual fixations is obvious, the most common being, everything goes my way, where improbabilities assist me; or, everything goes against me. These are not neurotic but psychotic. The idea that I made a great discovery, by inventing a word, for example, is also common in dreams, especially under stress, as reported by many writers and poets. In both neurotic-like and psychotic-like dreams and day dreams there is expression of wish-fulfillment, straightforward or in conflict (e.g. nightmares), but the one fixates on symbols and the other being definitely coherent (quite paradoxically, of course). Finally dreams and day dreams can be neurotic and psychotic simultaneously or alternatingly. The wish expressed by psychotic dreams (not to confuse it with dreams of psychotic patients) is all too often the abandonment of all responsibility. Day dreams of depressives often express this wish most clearly. All this accords with Freud's wish fulfillment theory as well as with the demarcation of psychosis by Yellowlees, Szasz, and many others, as the tiredness (exhausion) from the burdens of responsibility. This also ties Baruk's criticism of Freud's theory as one according to which the sense of responsibility has no role in problems of mental health, with the criticism of Freud's view of

psychosis as a mere severe neurosis. See H. Baruk's Introduction to I. Dorion, *l'Homme Moïse,* Zikarone, Paris, 1972.

As to the neurotic's attitude to responsibility, if he has one, it is not the tired desire to relinquish it. It may even be a desire to run the lives of others. Temper tantrums are the standard instance. Here we see that the megalomania of the psychotic is so very different from that of the (neurotic) authoritarian personality so-called. Also a neurotic may be afraid of resuming all responsibility such as Kafka's heroes are or Sartre's masochists. For Kafka the trial or the court is a symbol standing for a fear of deciding and is neither a fixation nor an abstract idea, and so not psychotic (perhaps also not neurotic). Sartre's story of sado-masochism (everyone is a sadist and a masochist) is a Freudian variant of Hegel's theory of the master-slave relation (every two people are a master and a slave) and obviously false. But a neurotic authoritarian person may terrorize a colleague or a spouse who is afraid to decide, be he a psychotic (depressive) or not. The patterns are clearly different in these two cases, with the depressive spouse being systematic and pseudo-rational, whereas the other, especially the masochist, more hysterical. Again the Kafka type indecisions may be impossible to fit either category – as a type both of people and of daydreams.

See also R. Fleiss, *Symbol, Dream and Psychosis,* International Universities Press, New York, 1973 (*Psychoanalytic Series,* vol. 3).

[28] K. Jaspers, 1913-1946, *General Psychopathology,* Manchester University Press, Manchester, 1963, p. 692-694, suggested we should apply the theory of process to the study of schizophrenia. The process is an organizational change (break) of the personality, and a lasting one; every further dissolution is a natural outcome of the process. This view of schizophrenia as a series of organizational changes was studied by Ey. But this theory is problematic since it makes the diagnosis of schizophrenia depend on a prognosis. Kraepelin's theory is similarly deficient. See Chapter 2, note 23 above.

[29] There is the traditional gradation from the normal to the severely pathological, where mild neurosis, severe neurosis, mild psychosis and severe psychosis, are in line. Here mild psychosis is more severe than severe neurosis. What is a mild psychosis is not clear. We may view ambulatory psychotics as suffering from mild psychosis which are, however, severer than a severe neurosis. Yet some neuroses are so severe as to require hospitalization. We may, alternatively, ignore hospitalization as hardly indicative of any degree of severity and find mild psychosis in cases of psychotic episodes (hospitalizable or not). Yet there are neurotic episodes, such as temper-tantrums and hysterical outbursts and hysterical epilepsy, which may be mild or severe. And so, the episodic nature is no criterion of severity, just as the periodic hospitalization of neurotics and of psychotics is not; it happens in both mild and severe cases. Hence we do not know what is a mild psychosis. Nor why neurosis is milder than psychosis. Nor is it clear how severe a condition has to be before the patient 'breaks down,' i.e. declares his own insufficiency or inability to cope, in one way or another.

The very case of breakdown or of insufficiency is, in itself, a matter of a patient's view of himself and decision about himself. (As Ey says, any mental disorder is an illness of freedom).

We may declare any breakdown psychotic, though obviously, the condition of it may be neurotic. For, we do have hysterical attacks, temper tantrums, etc. Yet, in our view, the very judgment and decision of every patient, his very act of breakdown, is psychotic. See next three notes.

[30] The existence of spontaneous remission may well indicate the fact that the psychotic patient keeps underground and working a whole system of thought similar to our public one and richer.

[31] That paranoia is a thought defect solves another and most central paradox: the paranoic breakdown, and generally the entering into the paranoic state, is both foolish and voluntary. This looks odd because the voluntary is deliberate and so rational, except on the supposition that the patient corners himself, and is either emotionally and intellectually exhausted, and so ready for the sought-for breakdown, or intellectually in a rut, which is a fixation.

[32] Ey sees all mental illness as illness of freedom. See H. Ey, *La Conscience,* Presses Universitaires de France, Paris, 2nd ed., 1968. He suggests that there is place for a psychopathology of the axiological system of the person. An English translation (by J. Flodstrom) is due from Indiana University Press.

See Y. Fried and F. Brüll, 'Intensive Psychotherapy for Acute Psychiatric Patients', *British J. Psychiatry* 121, 565; 635-639 (1972).

ANNOTATED BIBLIOGRAPHY

Abelson, R. P. and Rosenberg, M. I., 1958, 'Symbolic Psycho-logic: a Model of Attitudinal Cognition'. *Behav. Sci.* 3, 1-13.

For these authors psychological behaviour reflects logical underlying structures. McLaughlin (1963) (see below), based his interesting hypothesis on Abelson and Rosenberg's article.

Ackerknecht, E. H., 1957, *A Short History of Psychiatry,* Hafner, New York, 1959.

"The problem is one of trying to understand and classify symptoms, not causes in the material sense." (Crowcroft, 1967, p. 15).

Agassi, J., 1971/2, 'The Twisting of the I.Q. Test', *Philosophical Forum* 3, 265-277.

Agassi, J., 1975, *Science in Flux; Boston Studies in the Philosophy of Science,* vol. 28, Reidel, Dordrecht and Boston.

Agassi, J., 1976, *Towards a Rational Philosophical Anthropology,* Van Leer Jerusalem Foundation, 1976.

Alexander, F. G. and Selesnick, S. T., 1966, *The History of Psychiatry,* Harper & Row, New York.

Psycho-analytically orientated.

Allport, F. H., 1955, *Theories of Perception and the Concept of Structure,* Wiley, New York.

Modifies Gestalt theory to say that even 'mere' perception depends on the subject's activity; yet Gestalten are not so much given as constructed.

Allport, G. W., 1954, *The Nature of Prejudice,* Addison-Wesley, Cambridge, Massachusetts.

Defines the cognitive approach conditioning prejudice as the one where the ethnic attitude (prejudice) is 'based upon a faulty and inflexible generalization".

Altschule, M. 1957, *Roots of Modern Psychiatry,* Grune & Stratton, New York.

Raises original issues in the history of psychiatry from unexpected points of view.

Arieti, S., 1955, *Interpretation of Schizophrenia,* Brunner, New York, new and enlarged ed., 1975.

Deals with the schizophrenic's thought structure, which is considered highly regressive, but still purposeful.

Bacon, F., 1889, *Novum Organum* (ed. by T. Fowler), Clarendon Press, Oxford.

This most classic book on method declares all jumping to conclusions an error and a sin against reason and the source of prejudice and superstition and controversy.

Bannister, D., 1968, 'The Logical Requirements of Research into Schizophrenia', *Brit. J. Psychiat.* 114, 181-188.

Argues that a demarcation of schizophrenia is sorely needed, since otherwise true descriptions of patients suffering from different maladies get confused.

Bartley, III, W. W., 1964, *The Retreat to Commitment,* Chatto & Windus, London.

An attempt to take the irrational critique of rationalism as seriously as possible and a subsequent attempt to present rationalism not as a commitment in any sense (since it may always be revoked).

Barton, R., 1959, *Institutional Neurosis,* Wright, Bristol, 2nd ed., 1966.

Symptoms and signs of chronic schizophrenia are often the result of long hospitalization.

Baruk, H., 1945, *Psychiatrie morale expérimentale individuelle et sociale: Haines et reactions de culpabilité* (2nd ed.), Presses Universitaires de France, Paris, 1950.

A very unusual view of Psychiatry: a spiritual-moral dimension has to be added to psychopathology; one which is fundamentally expressed in the Jewish tradition of justice.

Bastide, R., 1965, *Sociologie des maladies mentales,* Flammarion, Paris. (Transl.: *The Sociology of Mental Disorder,* Routledge and Kegan Paul, London, 1972.)

A promising title by an eminent French professor of sociology. Disappointing. Whereas anti-psychiatric sociologists, though sometimes wrong (or vexing), may at least make interesting and even stimulating reading, Bastide is flat, tiresome, and unoriginal.

Bateson, G., Jackson, D. D., Haley, J., and Weakland, J., 1956, 'Towards a Theory of Schizophrenia', *Behav. Sci.* 1, 251-264.

This is, for some obscure reason, a classic paper. It claims that schizophrenia results from the double-bind, i.e., from choices imposed on the individual, where each alternative is wrong. This should render schizophrenic all of us, or at least all neurotics and all psychotics.

See Chapter 8, notes 8 and 9.

Benedetti, G., 1964, *Der Psychisch Leidende und seine Welt,* Hippokrates Verlag, Stuttgart.

The *verrückt* world – the spectacles make their user paranoic (L. Binswanger).

Binswanger, L., 1933, *Über Ideenflucht,* Miehans, Zürich.

Binswanger, L., 1960, *Melancholie und Manie,* Naske, Stuttgart.

See notes 19, 20 and 21 to Chapter 9 above.

Bion, W. R., 1959, 'Attacks on linking'. *Int. J. Psycho-anal.* 40, 308-315.

Bion, a Kleinian psychiatrist, attempts to understand the defective intellectual functioning in schizophrenia, as stemming from a defective unresolved emotional problem. The schizo (split) in the sphere of affect is the origin of the schizo (split) in the sphere of intellect (and language).

Biran, S., 1960, 'Versuch zur Psychopathogenese der Schizophrenic', *Psychiatria, Neurologia, Neurochirurgia (Netherlands)* 63, 252-281.

See Chapter 2, note 18.

Blanc, C., 1957, 'Neurobiologie et psychiatrie', *Evol. Psychiat.* 4, 625-652.

A neuropsychiatrist and a philosopher, Blanc calls for the union of phenomenology and genetic psychology (Piaget).

Blaney, P., 1975, 'Implications of the Medical Model and Its Alternatives', *Am. J. Psychiat.* 132, 911-914.

Just as we cannot blame the epileptic for his spasm, we cannot blame the madman for his folly, since both are physical conditions. By the same token, let us add, materialists have to ignore all moral considerations, which is absurd. Nevertheless, the author does use this line of thought to import into the Establishment stronghold some softer versions of the anti-psychiatrists, such as C. Rogers.

Bleuler, E., 1911, *Dementia Praecox oder die Gruppe der Schizophrenien,* in Aschaffenburg, G. (ed.), *Handbuch der Psychiatrie,* Spezieller Teil, Deuticke, Leipzig und Wien. English transl. by Zinkin, I., *Dementia Praecox or the Group of Schizophrenias,* International University Press, New York, 1950.

Schizophrenia in the light of Jackson's principles. Clinical pictures of the disease as a *mélange* of primary and secondary symptoms and signs; a classic and a real breakthrough in the history of psychiatry.

Bleuler, E., 1924, *Textbook of Psychiatry* (English transl. Brill, A.), Macmillan, New York, *Lehrbuch der Psychiatrie,* Bleuler, M. (ed.), 1966, Springer, Berlin.

Bolgar, H., 1964, 'J. Piaget and H. Hartmann: Contributions Towards a General Theory of Mental Development', in J. Masserman (ed.), 1964, *Science and Psychoanalysis,* vol. III, Grune & Stratton, New York.

Bonnafé, L., Ey, H., Follin, S., Lacan, J., and Rouart, J., 1950, *Le problème de la psychogenèse des névroses et des psychoses,* Desclée de Brouwer, Paris.

Leading French psychiatrists keenly discuss the problems of psychogenesis and its role in psychiatry.

Brand, J., 1971, *see* Mora, G., and Brand, J. (1971).

Breuer, J. and Freud, S., 1893-1895, *Studies on Hysteria,* Hogarth Press, London, *S.E.* 2 (1955), repr. 1957, 1962, 1975.

Both Breuer and Freud, in accord with the conventions and traditions of their epoch, permit psychotic states to be a constitutional factor if they occur as parts of hysteria. Later Freud altered his views a few times.

Brüll, F., 1969, 'The Trauma: Theoretical Considerations', *Israel Annals of Psychiat.* 7, 69-108.

A trauma is not anything traumatic that took place in an early past. Rather, a trauma is that which the analyst in retrospect views as significant. This makes the Freudian theory of the trauma always the means of being wise after the event and hence a definition rather than a hypothesis proper.

Brun, R., 1951, *General Theory of Neuroses,* International Universities Press, New York.

A psychoanalyst writes about the psychoanalytic theory of the neuroses in a biological language, without, however, reducing psychology to biology. The attempt reminds one of the early Freud, who formulated his psychology in terms of 19th century physics.

Brunschvicg, L., 1922, *L'expérience humaine et la causalité physique,* Alcan,

Paris.

Causality is both objective and an organizational principle of the perceiver. Brunschvicg was the teacher of both E. Meyerson and Piaget. He viewed the intellectual *élan* as creative and as hence unpredictable, thus rendering it strikingly akin to Bergson's *élan vitale;* yet he was a staunch rationalist. His commentators are understandably often baffled.

Butler, Samuel, 1906, *Erewhon.*

The first book on antipsychiatry. In accord with social Darwinism physical ill health is deemed criminal in *Erewhon,* the way mental ill health is deemed in the west. And vice versa.

Cameron, N., 1963, *Personality Development and Psychopathology: a Dynamic Approach,* H. Mifflin, Boston.

See also his contributions to Arieti, S. (ed.), *American Handbook of Psychiatry,* Basic Book, New York, 1959, Vol. I, pp. 508-539, new edition, 1974, Vol. III, pp. 676-693, and to Freedman, A. and Kaplan, H. (eds.), *Comprehensive Textbook of Psychiatry,* Williams & Wilkins, Baltimore, 1967, pp. 665-675.

The paranoic lives in a 'pseudo-community': he denies the objective community in which he lives and creates an imaginary (generally, persecutory) one instead. See our Chapter 5 note 11.

Capgras, J., 1909, *see* Sérieux, P. and Capgras, J. (1909).

Chambers, R., 1959, *see* Wotton, B., 1955.

Chapman, J. P., 1973, *see* Chapman L. J. and Chapman, J. P. (1973).

Chapman, L. J. and Chapman, J. P., 1973, *Disordered Thought in Schizophrenia,* Meredith Corp. (Appleton-Century-Crofts), New York.

A review, bibliography, and an excellent methodological criticism of the present situation in the research of schizophrenic thinking.

Clausen, J. A., 1957, *see* Leighton, A. H. *et al* (1957).

Crowcroft, A., 1968, *The Psychotic: Understanding Madness,* Pelican, Harmondsworth.

Psychosis, not only neurosis, is permitted in everyday life.

Dalbiez, R., 1936, *La méthode psychanalytique et la Doctrine Freudienne.* (2nd ed.), Desclée de Brouwer, Paris, 1949. English trans. Longmans Green, New York, 1941.

A non-orthodox psychoanalytical study synthesising Freud and Pavlov.

Duhem, P., 1906, *The Aim and Structure of Physical Theory,* Princeton University Press, 1954.

The most outstanding exposition to date of the conventionalist-instrumentalist view of science. Considering a theory as informative, we must admit that it is most unlikely to be true. Preferably, we should view theoretical system is always what is its domain of application. When extended beyond tive. The role of such systems is that of language-systems, namely, of classifying facts neatly conveniently and usefully. The question regarding a system is always, what is its domain of application. When extended beyond that domain, a system becomes cumbersome and calls for a modification. To this Duhem added his continuity principle: all modifications are made step by small step.

Durkheim, E., 1895, *The Rules of Sociological Method,* The Free Press, New York, 1958.

Durkheim, E., 1897, *Suicide: A Study in Sociology,* The Free Press, New York, 1951.

Gives the classical definition of deviation from a sociological viewpoint.

Needless to say, from our viewpoint, his definition of anomie signifies most. Anomie is often taken as a synonym for alienation, which is a mistake: The former belongs to the odd man out, whereas the latter can characterize a whole sub-culture or sub-group etc.

Dyer, A. R., 1974, 'R. D. Laing in Post-Critical Perspective', *Brit. J. Psychiat.* 124, 252-259.

Laing seems deviant and thus, to apply his view of deviants to himself, he seems mad. To make him look less mad we can compare his views to those of others, especially to those of Merlau-Ponty and Michael Polanyi, but also to views of some traditional philosophers and even to fairy tales.

Is it a compliment or an insult to call Laing mad?

Ellenberger, H. F., 1970, *The Discovery of the Unconscious: The History and Evolution of Dynamic Psychiatry,* Basic Books, New York.

Indispensable for the comprehension of the *Zeitgeist* influencing such people as Mesmer, Charcot, Janet and Freud in their theoretical understanding of mental illness both as disease and a mental phenomenon.

Encyclopédie Médico-Chirurgicale: Psychiatrie, 1955, 3 vols., Encyclopédie Médico-Chirurgie Publishers, Paris, 18 Rue Séguier (Paris 6e).

Erikson, E. H., 1964, *Insight and Responsibility, Lectures on the Ethical Implications of Psychoanalytic Insight,* Norton, New York.

An attempt to show that Freudianism is consistent with the view of man as responsible.

Erikson, E. H., 1969, *Identity: Youth and Crisis.* Norton, New York (Austen Riggs Monograph No. 7).

Identity as a stage in the process of normal healthy maturation is reached when the individual finds his proper role in society. From here to conformity and 'adjustment' is a stone's throw.

Esquirol, J. E. D., 1838, *Des maladies mentales considérées sous les rapports médical, hygiénique et médico-légal,* 2 vols., Baillière, Paris, also Tircher, Bruxelles, 1838. (Transl. by Hunt, E. K., *Mental Maladies: A Treatise on Insanity,* Lea and Blankhard, Philadelphia, 1845; Hafner, New York, 1965.)

Esterson, Aaron, 1970, *The Leaves of Spring: Schizophrenia, Family and Sacrifice,* Tavistock, London.

The family decides to cause the patient his illness. From this often true observation anti-psychiatry is somehow deduced.

Evans, P., 1972, 'H. Ey's Concepts of the Organization of Consciousness and Its Disorganization: an Extension of Jacksonian Theory', *Brain* 95 (part II), 413-440.

Evans-Pritchard, E. E., 1950, *Social Anthropology and other Essays,* The Free Press, New York, 1962.

The author shows how magic, mystical belief, etc. are part of everyday life, how they form an integrative system of thought, given that one understands it as the background of the social structure of a given society.

Ey, H., 1948-1954, *Etudes psychiatriques,* 3 vols., Desclée de Brouwer, Paris.
Ey's unmatched studies of psychopathology are presented here. An extension of Jackson's principles into psychopathology.

Ey, H., 1950, *see* Bonnafé, L. *et al.* (1950).

Ey, H., 1952, 'Grundlagen einer organo-dynamischen Auffassung der Psychiatrie', *Fortschr. Neurol.* 20, pp. 195 ff., reprinted in Strauss, E., and Zutt, J., (1963).

Ey, H., 1953, *Les délires (généralités), cours (revu 1967), Mimeograph.*
The first rendering of the paradoxes of paranoia (the critical analysis of the concept of delusion).

Ey, H., 1959, 'Unity and Diversity of Schizophrenia', *Amer. J. Psychiat.* 115, 706-714.
Ey suggests looking upon all psychoses as one genus, and upon the different psychoses as being species of this genus. His demarcation of psychosis is as a lower level of organization of the individual; this is also what defines the individual as sick. This resolves the paradoxes, we think.

Ey, H., 1962, 'Hughlings Jackson's Principles and Organo-Dynamic Concept of Psychiatry', *Amer. J. Psychiat.* 118, 673-682.
An overview of Ey's psychopathology. Jackson's principles are conceived in a generalized form, so as to cover psychiatric phenomena. This is Neo-Jacksonism; it was developed by Ey in Ey and Rouart (1936), and a further development thereof is to be found in Ey (1948-1954) and Ey (1974).

Ey, H., 1973, *Traité des hallucinations,* 2 vols., Masson, Paris.
This is Ey's *magnum opus.*

Ey, H., 1975, *Des idées de Jackson à un modèle organo-dynamique en psychiatrie,* Privat, Toulouse.

Ey, H., and Rouart, J. 1936, 'Application des principes de Jackson à une conception dynamique de la neuro-psychiatrie', *Encéphale* 31, 313-356; 31, 30-60, 96-123.
The first 'neo-Jacksonian' work. Published also as a monograph containing a French translation of H. Jackson's Croonian Lectures and of his Factors of Insanities. (*Selected Writings of J. H. Jackson,* ed. by J. Taylor, Hodder & Stoughton, London, 1931-1932). The monograph includes an Introduction by H. Claude. Doin, Paris, 1938.

Faberga, H., 1974, *Disease and Social Behaviour: An Interdisciplinary Perspective,* M.I.T., Cambridge, Mass.

Faberga, H., 1975, 'The Position of Psychiatry in the Understanding of Human Disease', *Arch. Gen. Psychiat.* 32, 1500-1512.
This concise paper contains a broadly drawn description, historical, biological, social, functional, philosophical, of disease in general and of mental illness in particular. We do not find ourselves qualified to comprehend it. Yet, since he says (p. 1511, final sentence) "A basic premise is the following: diseases may be viewed as social categories that are grounded in biocultural facts about man, and that also reflect man's judgment about his adaption,

etc.", we suspect that we strongly disagree with him.

Federn, P., 1952, *Ego-Psychology and the Psychoses,* Basic Books, New York, 1952; Imago, London, 1955.

The posthumous republication of the much delayed papers by the father of existential psychiatry – due to suppression by the Father. Schizophrenia is the loss of distinction between self and others due to loss of bodily boundary feelings.

Fenichel, O., 1946, *The Psychoanalytic Theory of Neurosis,* Routledge & Kegan Paul, London, 1966.

Following Freud, Fenichel views the psychopathology of psychosis, as essentially the same as that of a neurosis. The difference, he says, is in degree, intensity, etc. The gradation of the continuum Normal-Neurotic-Psychotic could not be better expressed.

Finnegan, R., 1973, *see* Horton, R. and Finnegan, R. (1973).

Follin, S., 1950, *see* Bonnafé *et al.* (1950).

Foucault, M., 1961, *Histoire de la folie à l'âge classique,* Gallimard, Paris, new ed. 1972. (Transl. of the first ed. *Folie et déraison,* Plon, Paris, 1961, as *Madness and Civilization: A History of Insanity in the Age of Reason,* Pantheon, New York, 1965.)

This work evolved out of Foucault's doctoral dissertation, a history of madness rather than a history of psychiatry proper, which was accepted by Jean Hyppolite, and also by G. Canguilhem, whose own 1943 medical dissertation concerns *Le normal et le pathologique* (Presses Universitaires de France, Paris, 1966) where he tries to demarcate one from the other.

It is *Histoire de la folie,* plus the political events of Paris, May 1968, that gave birth to the French version of anti-psychiatry.

Frank, G., 1975, *Psychiatric Diagnosis: A Review of Research,* Pergamon, Oxford.

This work is remarkable chiefly for its extensive bibliography, pp. 83-125. Its first 81 pages are fairly descriptive of the literature and of the state of the art. The bibliography is fairly extensive, but confined strictly to the English language.

Frankenstein, C., 1966, *The Roots of the Ego,* Livingstone, Edinburgh. *See* note 10 to Chapter 7.

Freud, Anna, 1936, *The Ego and the Mechanism of Defence,* International Universities Press, New York, 1946, revised 1966; Hogarth Press, London, 1948.

This is the classic introduction of the concept of defence-mechanism which is now so much an integral part of both Freudian theory – it is only implicit in S. Freud – and in everybody's everyday life. As Rapaport, D. (1960), notes (p. 208) it is both a schema and a meta-concept, relating mainly to communication barriers. It is therefore no surprise that S. Freud did not present it explicitly even though it is a central integrative principle of his own system that does not break its seams. Freud himself, however, suggested that neurosis *also* is the most economical solution (a note added in 1923 to his [1901], 1905, p. 43; also in Chapter 3 to his 1926), and thus we may add, it also defends one from reverting to psychosis; what, then, does psychosis defend one from reverting to? Suicide, of course. Freud considered the absence of all defences the (impossible) normal; yet he should have also endorsed as the (impossible) normal a perfect defence system. This is what, we suggest, the paranoic aims at, and almost achieves. But this way, the idea of defence mechanisms does break the seams of the system.

Freud, S., 1891, *On Aphasia, a Critical Study* (ed. Stengle, E.). International Universities Press, New York, 1953.

In this monograph Freud largely uses H. Jackson's principles. In his further works Freud scarcely ever mentions Jackson's name.

Freud, S., 1893-1895, *see* Breuer, J. and Freud, S. (1893-1895).

Freud, S., 1895, *Paranoia (Draft H from the Fliess Papers),* Hogarth Press, London, *S.E.* 1 (1975) pp. 206-212.

Freud's first, in essence best, on paranoia. Paranoia as a defence mechanism already then!

Freud, S., 1900, *The Interpretation of Dreams,* vols I and II. Hogarth Press, London, *S.E.* 4 and 5 (1975).

Freud, S., 1901, *Psychopathology of Everyday Life,* Hogarth Press, London, *S.E.* 6 (1960), repr. 1975.

Freud, S. [1901], 1905, *Fragment of an Analysis of a Case of Hysteria,* Hogarth Press, London, 1975, *S.E.* 7, pp. 7-122.

Freud, S., 1916-1917, *Introductory Lectures on Psychoanalysis,* Hogarth Press, London, *S.E.* 15 and 16 (1975).

The pathology of everyday life is portrayed, as well as dreams, so paving the way for the construction of a general theory of neurosis. In so doing Freud humanizes mental illness, and makes it comprehensible as a common and an almost normal human experience.

Freud, S., 1911, *Psychoanalytic Notes on an Autobiographical Account of a Case of Paranoia (Dementia Paranoides),* Hogarth Press, London, *S.E.* 12 (1975), 3-82.

Freud relates paranoia to repressed (latent) homosexuality. He describes paranoic thinking processes as essentially a projection, built upon 3 articulations: Assertion: "I love him"; Denial: "I hate him"; Reversal: "He hates me". The reversal is, of course, a projection. (Hence, projection is merely derivative!)

Freud, S., 1923, *The Ego and the Id,* Hogarth Press, London, *S.E.* 19 (1961), repr. 1975, pp. 3-66.

The classic mapping of the mind into three parts. Notice that the Id or unconscious is inborn, whereas the subconscious includes the super-ego which is repressed. Freud bitterly laments the harshness of our moral code which pushes people all too often to despair and even into a life of crime.

Freud, S., 1926, *Inhibitions, Symptoms and Anxiety,* Hogarth Press, London, *S.E.* 20 (1975), pp. 87-156.

Freud's important discussion of primary and secondary gains. The nearest he comes to Jackson, when not mentioning him.

Freud, S., [1938], 1940, *An Outline of Psychoanalysis,* Hogarth Press, London, *S.E.* 23 (1975), 141-207.

Freud's later view on psychoanalysis. Here the theory of fixation is formulated in such a way that it becomes clear that in his opinion it is in adolescence that the individual either stays with or shakes off his early fixations; and thus becomes either neurotic or psychotic or grows up as a healthy person.

Fried, Y., 1954, 'De la régression intellectuelle chez des schizophrènes', Report presented to J. Piaget, Geneva. *Rivista di Psicologia Sociale, e Archivio Italiano di Psicologia Generale e del Lavoro* (Torino) 1 (1956), 53-69; 2, (1956), 83-103.

The existence of an egocentric form of thinking (in the sense of Piaget) in paranoid schizophrenic patients is suggested as an explanation of the clinical symptom. Hallucination is viewed as a necessary derivative of an egocentric form of thinking. Or, if you will, children in the egocentric stage quite normally hallucinate.

Fried, Y., 1973, 'Thinking in Paranoia: a Comparative Study of Conceptual Models', *Brit. J. Med. Psychol.* 46, 347-358.

Paranoia does not consist of a limited disturbance represented only by a delusion, but rather reflects a general organization of thinking of a particular structural type. A detailed description of a clinical case study of a paranoic patient follows. A conceptual model is proposed, according to which paranoic thinking is one track at a time.

Fried, Y., 1974, 'The Methodology of the Odd Occurrence in Clinical Medicine and research', *Agressologie* 15, 5-10.

Following an analysis of Claude Bernard's methodology of the counter-expected occurrence in medicine, it is suggested that two difficulties are in this method: an intellectual difficulty, and an emotional one. These difficulties hinder the clinician from using the method though the counter-expected is quite frequently met in daily work. To remedy this in medical clinical practice and research, the teaching of methodology in medical schools is recommended, including the study of psychological overtones of research.

Fried, Y., 1975, 'Jacques Lacan's Conception of Psychoanalysis', *Contemporary Psychoanalysis* pp. 251-254.

Psychoanalysis begins where the patient say anything whatsoever to his psychoanalyst. We may view Jacques Lacan's conception of psychoanalysis as a kind of calculus, where layers of interpretation are unravelled.

Fromm-Reichmann, Frieda, 1959, *Psychoanalysis and Psychotherapy. Selected papers.* Edited by D. M. Bullard, Foreword by Edith V. Weigert, University of Chicago Press.

The first paper discusses the facts of the paradoxes of paranoia and includes even such details as the psychotic's aversion to accepted hypocrisy. Claims that transference of sorts with psychotics is quite possible, and so even psychoanalysis of sorts.

Gelner, E., 1973, *see* Horton, R., and Finnegan, R. (eds.), 1973.

Ginsberg, M., 1947, 'Anti-Semitism', pp. 196-212, in his *Reason and Unreason in Society: Essays in Sociology and Social Philosophy.* Longmans Green, London, new impression, 1948.

A discussion of the carrier of a prejudice which strongly resembles Kraepelin's paranoic. See Chapter 3, notes 7, 9, 10 and 13 above.

Gluckman, M. (ed.), 1964, *Closed Systems and Open Minds: the Limits of Naivety in Social Anthropology,* Aldine, Chicago.

Goffman, E., 1961, *Asylums,* Doubleday/Anchor, New York.
Asylums viewed as a total institution are compared to the military, monasteries, and small colleges. The structural hierarchy of the mental hospital is described, where the patient is at the lowest level.

Gouin-Décarie, T., 1962, *Intelligence and Affectivity in Early Childhood: An Experimental Study of Jean Piaget's Object Concept and Object Relations.* International Universities Press, New York, 1965.
The author, a psychoanalyst and a Piagetian uses empirical findings to delineate a parallelism between Freud's view of the affective side in the development of young children, and Piaget's view of their intellectual development. There is no evidence of any predominance of the one or the other of these two aspects of behaviour. Piaget contributed a very stimulating introduction to Gouin-Décarie's study, in which some problems of psychopathology are discussed. This is significant as the only place where the extensions of his theory to psychopathology are mentioned by him.

Goldstein, K., 1940, *Human Nature in the Light of Psychopathology,* Harvard University Press, Cambridge, Mass. 1951.
See Chapter 7, note 2, and Chapter 9, note 21.

Goldstein, K. and Scheerer, M., 1941, 'Abstract and Concrete Behaviour: an Experimental Study with Special Tests', *Psychological Monographs,* 239, 1-151.
A classic work describing the 'concrete thinking' in schizophrenia. We deem our indebtedness to this study obvious. See Appendix I and Chapter 9, note 22.

Gourwitsch, A., 1957, *Theorie du champ de la conscience,* Desclée de Brouwer, Paris.

Gruenberg, E. M., 1957, 'Socially Shared Psychopathology', pp. 201-229, in Leighton, A. H., Clausen, J. A. and Wilson, R. N. (eds.), *Explorations in Social Psychiatry,* Basic Books, New York.
See our annotation to Leighton, A. H. *et al.* (1957).

Haley, J., 1956, *see* Bateson, G. *et al.* (1956).

Hart, H. L. A., 1968, *Punishment and Responsibility, Essays in the Philosophy of Law,* Oxford University Press, New York & Oxford.
An important discussion of M'Naghten's rules.

Hartmann, H. 1953, *Essays on Ego Psychology, Selected Problems in Psychoanalytic Theory,* International Universities Press, New York, 1964. Chapter 10, 'Contribution to the Metapsychology of Schizophrenia'.
The author operates with the cathexis model, which we reject with most writers, and with many Freudians, as pseudo-scientific; the author modifies Freud's model, thus adding material we find incomprehensible; we find it impossible to comment on his work. See, however, the higly favorable review by Holt, R., 1959, *Contemporary Psychology* 4, 332-333.

Henry, G., 1941, *see* Zilboorg, Z., and Henry, G. (1941).

Hesnard, A., 1949, *L'universe morbide de la faute,* Presses Universitaires de France, Paris.
Guilt feelings discussed from a combined psychoanalytic viewpoint and a moralistic one.

Holt, R. (ed.), 1967, *Motives and Thought,* Psychological issues, vol. V (213), Monograph No. 18/19, International Universities Press, New York.
Some of D. Rapaport's disciples present studies of pathological behaviour from a cognitive and formal point of view. The interest here is precisely in that though the authors belong to the psychoanalytic school, they reject Freud's view of thinking as merely a byproduct of affect.

Horton, R., and Finnegan, R. (eds.), 1973. *Modes of Thought: Essays on Thinking in Western and Non-Western Societies,* Faber and Faber, London.
Horton distinguishes between open and closed systems, à la Popper, where closed systems are elastic but imprisoning, whereas open systems are less elastic and allow for competition. The present volume offers both variants and criticisms of this view. Since we think that methodologically, paranoia is somewhere in between what Horton calls closed and open systems, we think the reader may benefit from perusing this volume: we recommend, especially, Ernest Gellner's contribution, 'The Savage and the Modern Mind', pp. 162-181, which is a vivid critical analysis of Horton and an excellent introduction to the topic of its title.

Inhelder, B. and Piaget, J., 1955, *De la logique de l'enfant à la logique de l'adolescent: Essai sur la construction des structures opératoires formelles.* Presses Universitaires de France, Paris (Transl. *The Growth of Logical Thinking from Childhood to Adolescence,* Routledge & Kegan Paul, London, 1958; Basic Books, New York, 1958.)

Jackson, D. D., 1956, *see* Bateson, G. *et al.* (1956).

Jackson, J. Hughlings, *Selected Writings of J. Hughlings Jackson* (ed. by Taylor, J.), 2 vols., Hodder & Stoughton, London, 1931-1932.

Jarvie, I. C., 1965, 'Limits to Functionalism and Alternatives to it in Anthropology', in Martindale, D. (ed.), *Functionalism in the Social Sciences,* No. 5 in a series of monographs sponsored by the American Academy of Political and Social Science; repr. in Manners, R., and Kaplan, D. (eds.), *Theory in Anthropology: A Source Book,* Aldine, Chicago, 1968, pp. 196-203.
A scathing attack on the poor logic of functionalist thinking in anthropology.

Kaplan, B., 1963, *see* Werner, H. and Kaplan, B. (1963).

Kaufmann, W., 1973, *Without Guilt and Justice: From Decidophobia to Autonomy,* Wyden, New York.
Introduces the new term, decidophobia, to denote a normal situation which, however, is so common in psychopathology.

Klein, M., 1921-1945, *Contributions to Psychoanalysis,* Hogarth Press, London, 1948.
The seeds of psychosis are present in the ego of the newborn infant.

Kline, P., 1972, *Fact and Fantasy in Freudian Theory,* Methuen, London.

A critical review of psychoanalysis, inspired by Eysenck's views. Too much statistically minded for our taste.

Koestler, A., 1959, *The Sleepwalkers,* Grosset & Dunlap, New York.

Kohut, H., 1971, *The Analysis of Self, A Systematic Approach to the Psychoanalytic Treatment of Narcissistic Personality Disorder,* The Psychoanalytic Study of the Child, Monograph No. 4, International Universities Press, New York.

If hypochondria and schizophrenia are both narcissistic, how come they are not both psychotic? Because the archaism of the former is isolated and a mere fragment, whereas the latter is a global disturbance and so properly psychotic. Embarrassingly orthodox Freudian, taking seriously Freud's ideas that narcissism prevents the psychotic's transference. Hence, all observed narcissist's transferences are not true blue: they are called "mirror transference", as opposed to Freudian "idealizing transference".

Kraepelin, E., 1883-1927, *Kompendium der Psychiatrie* (1883*); *Psychiatry* (1910*, 8th ed., 1927; 9th ed., Vol. I) *Allgemeine Psychiatrie,* J. Lange, Vol. II, *Klinische Psychiatrie* (Kraepelin) Vols. II & IV, new impression of the 8th ed., Barth, Leipzig.

See Chapter 2, note 5.

An abbreviated translation of the 6th ed. of Kraepelin's textbook is available in English: Kraepelin, E., *Textbook of Psychiatry* (trans. by A. R. Defendorf – abstracted) Macmillan, London, 1902. See R. O. Rieder, 'The Origins of Our Confusion about Schizophrenia', *Psychiatry* 37 (1974), 197-208.

Portions of Kraepelin's textbook, 8th ed., are available in English translations in two books: E. Kraepelin, *Dementia Praecox and Paraphrenia,* Livingstone, Edinburgh, 1919; E. Kraepelin, *Manic-Depressive Insanity and Paranoia,* Livingstone, Edingburgh. Both books were translated by R. M. Barclay. *Nerv. and Ment. Dis. Monog.* 14, is the chapter *General Paresis.* Kraepelin's *Lectures on Clinical Psychiatry,* 1900, appeared in English (tr. Johnstone, T.), from Baillière, London 1904.

Kruse, H. D. (ed.), 1957, *Integrating the Approaches to Mental Diseases,* A Symposium of the New York Academy of Medicine, Hoeber-Hayer, New York.

Outstanding for its participants' balanced and serious handling of the very complex matter of psychiatric schools. To this publication of 1957, one should add the newer views of Foucault, Szasz and Laing. Otherwise it all stands.

Lacan, J., 1932, *De la psychose paranoïaque dans ses rapports avec la personnalité,* Le François, Paris; New enlarged ed., Ed. du Seuil, Paris, 1975.

Contra Jaspers: Paranoia is not unrelated to the personality, nor a break of the personality.

Lacan, J., 1950, *see* Bonnafé, L. *et al.* (1950).

Lacan, J., 1966, *Ecrits,* Ed. du Seuil, Paris.

Lacan, J., *The Language of the Self* (translated with notes by Wilden, A.), Johns Hopkins Press, Baltimore, 1968.

Laing, R. D., 1960, *The Divided Self,* Tavistock, London. Also Pelican, Harmondsworth (Paperback), 1965.

Laing, R. D., 1967, *The Politics of Experience and the Bird of Paradise,*

Penguin, Harmondsworth.
The locus classicus and the manifesto of anti-psychiatry.

Lemert, E., 1951, *Social Pathology: A Systematic Approach to the Theory of Sociopathic Behaviour,* McGraw Hill, New York.
See Chapter 3, note 4.

Leighton, A. H., Clausen, J. A., and Wilson, R. N. (eds.), 1957), *Exploration in Social Psychiatry,* Basic Books, New York.
This book obviously should raise the question, is mass psychosis possible? It comes to it in Part II, middle chapter (Ch. VII), on 'Socially Shared Psychopathology' by Ernest M. Gruenberg. It is evasive in the extreme: is the psychopathology it notices neurotic or psychotic? No hint at the question, let alone the answer.

Lévy-Strauss, C., 1962, *The Savage Mind,* Weidenfeld & Nicholson, London, 1966.
This celebrated work offers the general idea that primitive science is the science of the concrete, where symbol is identical with the thing symbolized, whereas in modern science the distinction between symbol and its meaning is obvious. The modern science of communication is the basis of our understanding of savage symbols (e.g. totems) to be just that.

Lewis, A., 1963, 'Medicine and the Affections of the Mind', *Brit. Med. J.,* 2, 1549-1557, reprint in Lewis, A., *The State of Psychiatry: Essays and Addresses,* Routledge & Kegan Paul, London, 1967.
"The dependence of the majority of psychiatrists upon Freudian Psychopathology recalls the dominance of systems in medicine in the eighteenth century. . . . Popper regards it as a pre-scientific metaphysical scheme" (p. 285). Lewis continues to quote Popper (1963), thus: "the study of such theories seemed to have the effect of an intellectual conversion or revelation, opening your eyes to a new truth hidden from those not yet initiated. Once your eyes were thus opened, you saw new confirming instances everywhere: the world was full of verifications of the theory. Whatever happened always confirmed it. Thus its truth appeared manifest" (p. 285).
Had Lewis replaced 'the theories' by 'the patients', he would have hit on the definition of paranoia.

Lewis, A., 1970, 'Paranoia and Paranoid: A Historical Perspective', *Psychological Medicine* 1, 2-12.

Lidz, T., 1968, *The Person,* Basic Books, New York.
Extension of A. Meyer's (1952) psychobiological model, to include in it the role of the family, as a fundamental means by which culture is transmitted.

Lomas, P., 1973, *True and False Experience, the Human Element in Psychotherapy,* Taplinger, New York.
Both a solid and a popular, clear exposition of the post-Freudian schools, including the existentialists.

Mann, H., 1969, *see* Siegler *et al.* (1969).

McLaughlin, H. G., 1963, 'Psycho-Logic: a Possible Alternative to Piaget's Formulation', *Brit. J. Educ. Psychology,* 33, 61-69.
Though intended to be an alternative to Piaget, we consider it a generalization of Piaget's system and a breakthrough. Perhaps the most important

contribution to Piagetian psychology by one who is not of the school of Geneva. Essential to the understanding of psychopathology of thinking.

Mayer-Gross, W., 1924, *Selbstschilderungen der Verwirrtheit: Die Oneiroide Erlebnisform,* Springer, Berlin. Partly trans. in Ey, H. (1948-1954), *Etudes psychiatriques,* Desclée de Brouwer, Paris, III (1954), 250-279.

Mayman, M., 1963, *see* Menninger, K. (1963).

Menninger, K. (with Mayman, M. and Pruyser, P.), 1963, *The Vital Balance, the Life Process in Mental Health and Illness,* Viking Press, New York, 1969.
An attempt to develop a unitary theory of psychosis.

Merton, R. K., 1957, *Social Theory and Social Structure,* The Free Press, New York, (rev. ed.).
See Chapter 7, note 20.

Meyer, A., 1952, *Collected Papers* (ed. by Winters, E.), John Hopkins Press, Baltimore, 4 vols.
Meyer was a pupil of both Forel and Jackson. He viewed mental illness not as disease entities but as bio-psychological reaction types to the environment, integrated at the cultural (symbolic) level.

Meyerson, E., 1907, *Identité et realité,* Alcan, Paris, (English translation: *Identity and Reality*, Dover reprint, N.Y.).
See Chapter 5, note 3.

Minkowski, E., 1927, *La schizophrénie,* Desclée de Brouwer, Paris, 1953.

Minkowski, E., 1933, *Le temps vécu,* Desclée de Brouwer, Paris, 1953.

Minkowski, E., 1966, *Traité de psychopathologie,* Presses Universitaires de France, Paris.
A sharp insight into *expérience vécue,* Space and Time.

Moore, M., 1975, 'Some Myths about "Mental Illness" ', *Arch. Gen. Psychiat.* 32, 1483-1497.
A systematic presentation of the arguments against the view of psychotics as mentally ill and a rejoinder. The most up-to-date of its kind, i.e. of the establishment's enlightened defense against the new critics.

Mora, G., 1965, 'The History of Psychiatry: a Cultural and Bibliographical Survey'. *The Psychoanalytic Review* 52 (2), repr. in *International J. Psychiatry* 2, 335-356, Discussion pp. 357-366.
A critical review of the historical literature in psychiatry. An abundant bibliography.

Mora, G. and Brand, J., 1971, *Psychiatry and its History; Methodological Problems in Research,* Thomas, Springfield, Ill.

Morel, F. (ed.), 1950, *Psychopathologie des délires.* Vol. I of the Proceedings of the International Congress of Psychiatry, Herman, Paris.
Views of contemporary psychiatrists concerning delusions. See Chapter 1, note 5.

Murphy, G., 1947, *Personality: A Biosocial Approach to Origins and Structure,* Harper, New York.

Murphy, J. M., 1976, 'Psychiatric Labeling in Cross-Cultural Perspective', *Science* 191, 1019-1028.

This is the most balanced study of comparative psychiatry, and it covers most of the literature on the topic. It compares only two cultures, one African and one Eskimo, and shows it is easy to translate terms for madness from the one the other. For our part, we think this only shows how undifferentiating popular diagnosis is – in any culture. Certainly we do not see how it leads to the author's organicist conclusions.

Niederland, W. G., 1974, *The Schreber Case: Psychoanalytic Profile of a Paranoid Personality,* Quadrangle, New York.

Schreber's case is presented in the light of biographical facts: Schreber's father was a tyrant in the garb of a disciplinarian; all Schreber's persecution visions were true.

Noy, P., 1969, 'A Revision of the Psychoanalytic Theory of Primary Process', *Int. J. Psychoanal.* 50, 155-178.

One of the more successful attempts to integrate Piaget and Freud.

Odier, C., 1947, *L'angoisse et la pensée magique,* Delachaux et Niestlé, Neuchatel et Genève.

An attempt to integrate elements of Piaget's early writings (*The Child's Conception of the World* (1926), into psychoanalytic theory. One of the first psychoanalysts to see that Piaget's findings may complement Freud's. The result, however, is rather disappointing.

Osmond, H., 1969, *see* Siegler, M. *et al.* (1969).

Osmond, H., 1974, *see* Siegler, M., and Osmond, H. (1974).

Parsons, T., 1951, *The Social System,* The Free Press, New York.

See Chapter 7, note 20.

Payne, R. W., 1966, 'The Measurement and Significance of Overinclusive Thinking and Retardation in Schizophrenic Patients', in Hoch, P. and Zubin, J. (eds.) *Psychopathology of Schizophrenia,* Grune & Stratton, New York, pp. 77-97.

Payne's central idea is that schizophrenics tend to correlate bits and pieces of concepts and information into a general principle, except that the principle makes no sense. No doubt this often happens; the claim that it always does is false. Moreover, since bold innovations, such as non-Euclidean geometry, seemed nonsense to most scholars at first, the paradox of paranoia is raised here.

Piaget, J., 1923, 'La pensée symbolique et la pensée de l'enfant', *Arch. de Psychol. de Genève* 18, 273-304.

A report presented by Piaget at a psychoanalytic congress, with Freud presiding and showing benevolent indifference to the then young man's comparison of the thought of a child as understood by Freud and Piaget. *See* Gouin-Décarie (1962).

Piaget, J., 1945, *La formation du symbole chez l'enfant,* Lelachaux & Niestlé, Neuchatel & Paris. (transl. *Play, Dreams and Imitation in Childhood,* Heinemann, London, 1951.

Imitation, play, dream, and symbol, are presented as successive levels of cognitive organization. See Gouin-Décarie (1962).

Piaget, J., 1949, *Traité de logique: Essai de logistique opératoire,* Colin, Paris.

Piaget's theory of the parameters characterizing the structure of thought is expressed here as a course in formal logic.

Piaget, J., 1953-1954, 'Les relations entre l'intelligence et l'affectivité dans le development de l'enfant', *Bull. Psychol. de l'Univ. de Paris* 7, 143-150, 346-361, 699-701.

Piaget, J., 1955, *see* Inhelder, B. and Piaget, J. (1955).

Pichot, P., 1949, *Les tests mentaux en psychiatrie,* Presses Universitaires de France, Paris.

A comprehensive review of existing tests.

Pinel, P., 1798, *Nosographie Philosophique ou la Méthode de l'Analyse Appliquée à la Médicine,* 3 vols., Brosson, Paris, 1810.

Pinel, P., 1801, *Traité médico-philosophique sur l'aliénation mentale, ou la manie,* Brosson, Paris, 1809. (transl. *A Treatise on Insanity,* Hafner, New York, 1962.)

These celebrated works are much more books on general methodology of scientific thinking in medicine than books of (clinical) psychiatry. Yet, of course, they are comprehensive as far as psychiatry of the day is concerned.

Poincaré, H., 1908, *Science et méthode,* Flammarion, Paris.

Polanyi, M., 1951, *The Logic of Liberty,* Routledge and Kegan Paul, and University of Chicago Press, London and Chicago.

Polanyi, M., 1958, *Personal Knowledge: Towards a Post-Critical Philosophy,* Harper, New York, 1964.

Polanyi's *magnum opus.* He thinks it is less important for doctors to define madness or to tell us the difference between genuine and hysterical epilepsy, than to know one when they see one, and to have a consensus on it. Unfortunately we are less sanguine about doctor's abilities than he is.

Polanyi, M., 1959, *The Study of Man,* University of Chicago Press, Chicago.

Popper, K. R., 1935, *The Logic of Scientific Discovery,* Hutchinson, London, rev. sec. ed., 1968.

Popper's first vintage and *magnum opus.* He characterizes science as systems of theories, each of which, taken as a whole, can, in principle conflict with an observation report. (In principle, since, if true, then it may in fact be irrefutable. But it must be refutable in principle.) He presents the problem of induction as, how can we conclude from particular observations about universal theories? or, to put it less formally, how can we gain theoretical knowledge from empirical experience? He answers, by refutations. Includes a theory of simplicity as the paucity of parameters.

Popper, K. R., 1945, *The Open Society and its Enemies,* Princeton University Press, Princeton, N. J., 1950.

Views the ability to take responsibility in the open society as a sign of maturity, yet also an unrelievable burden, "the cross of civilization". This, he says, explains the tendency of intelligent people "to return to the cave" via the closed society.

Popper, K. R., 1963, *Conjectures and Refutations: The Growth of Scientific Knowledge,* Routledge & Kegan Paul, London.

A classic collection of papers on science, metaphysics and society. Presents a view of science most congenial to Piaget's theory of learning: all learning, including science, is by trial and error.

Popper, K. R., 1972, *Objective Knowledge: An Evolutionary Approach,* Oxford University Press, Oxford.

Pruyser, P., 1963, *see* Menninger, K. (1963).

Rapaport, D. (ed.), 1951, *Organization and Pathology of Thought: Selected Papers,* Columbia University Press, New York.

Incorporates the compiler's own psychoanalytic interpretations of Piagetian and other texts and vice versa.

Rapaport, D., 1967, *The Collected Papers of David Rapaport* (ed. Gill, M.), Basic Books, New York and London.

Raskin, D., 1975, 'Bleuler and Schizophrenia', *Brit. J. Psychiat.* 127, 231-234.

A long overdue *mis à point* of the confusion in current psychiatric textbooks, about Bleuler's fundamental and accessory signs in schizophrenia. Unfortunately the author himself is still somewhat less sharp than we deem desirable, regarding Bleuler's primary and secondary signs. See note 23 to Chapter 2 above.

Rieder, R., 1974, 'The Origins of Our Confusion about Schizophrenia', *Psychiatry* 37, 197-208.

An excellent historical presentation of the core of present views of schizophrenia with an eye constantly on the paradox of paranoia.

Rosenberg, M. I. 1958, *see* Abelson, R. P. and Rosenberg, M. I.

Rotschild, F. S., 1968, 'Concepts and Methods of Biosemiotic , *Scripta Hierosolymitana* 20, 163-194.

A study using the system approach (whatever this may mean) to medicine in general and psychiatry in particular. The idea of biosemiotics is that all phenomena of life should be viewed as linguistic. Language is viewed as structurized in hierarchical levels, according to degrees of complexity.

Rouart, J., 1936, *see* Ey, H. and Rouart, J. (1936).

Rouart, J., 1950, *see* Bonnafé, L. *et al.* (1950).

Rush, Benjamin, *The Selected Writings of Benjamin Rush* (ed. by T. Dobson), Philosophical Library, New York, 1947.

The American counterpart of Pinel, the Tukes, etc.

Russell, B., 1910, *Mysticism and Logic,* Allen & Unwin, London.

Contains the classic essay, 'A Free Man's Worship'.

Russell B., 1928, *Skeptical Essays,* Norton and Allen & Unwin, New York & London.

Russell, B., 1948, *Human Knowledge: Its Scope and Limits,* Simon & Schuster, New York, 1967.

Rycroft, C., 1968, *A Critical Dictionary of Psychoanalysis,* Penguin, Harmondsworth, 1972.

A gem. Freud, Klein, and the existentialist are in evidence here, but the word "critical" in the title is amply justified.

Sartre, J. P., 1943, *Existential Psychoanalysis,* Gateway paperback ed., H. Regnery Cie., Chicago, 4th print, 1966. See also Sartre, J. P., 1943, Part 1, ch. 2, and Part IV, ch. 2 of his *Being and Nothingness,* Philosophical Library, New York, 1953; Methuen, London, 1956.

An extension of Freud's idea of self-deception both as severe repression and as an illusion as to one's own identity (in the sense discussed under Wheelis, 1958).

Sattes, H., 1963, *see* Strauss, E., and Zutt, J., 1963.

Schatzman, M., 1972, *Soul Murder: Persecution in the Family,* Allen Lane, London. See Niederland, W. G. (1974).

Scheerer, M., 1941, *see* Goldstein, K. and Scheerer, M. (1941).

Scheff, T. J., 1966, *Being Mentally Ill: A Sociological Theory,* Aldine, Chicago.

Scheff summarizes his view of labelling theory in Scheff, T. J., 1970, 'Schizophrenia as Ideology', *Schizoph. Bull.* 2, 15-19, reprint. in Brown, P. (ed.), 1973, *Radical Psychology,* Tavistock, London. *See* also note 17 to Chapter 6 above.

Schipperges, H., 1970, *Modern Medezin im Spiegel der Geschichte,* Thieme, Stuttgart.

A remarkable chapter on medicine in the Romantic period. The natural consequence of Romanticism is the extermination of the mentally ill by the Nazis.

Schreber, D. P., 1903, *Memories of my Nervous Illness* (transl. by MacAlpine, I. and Hunter, R.), Dawson & Son, London, 1955.

Schwartz, M. S., 1954, *see* Stanton, A. H. and Schwartz, M. S. (1954).

Schweitzer, A., 1913, *The Psychiatric Study of Jesus: Exposition and Criticism* (transl. and with an Introduction by Ch. R. Joy), Beacon, Boston, Mass., 1948, 1962.

This is slim work, evidently hastily prepared in order to meet formal requirements. It is written more or less from a Kraepelinian point of view, and so the diagnosis is heavily dependent on the question, was Jesus the son of God. This question Schweitzer declares scientifically illegitimate.

See Chapter 1, note 29.

Seal, V. G., 1959, *see* Wootton, B., 1955.

Segal, H., 1964, *Introduction to the Works of Melanie Klein,* Basic Books, New York.

A comprehensive, concise exposition, abundantly illustrated with the author's own clinical work.

Selesnick, S. T., 1966, *see* Alexander, F. G. and Selesnick, S. T. (1966).

Seligman, P., 1972, 'Review of E. Strauss, Natanson, M. and Ey, H., *Psychiatry and Philosophy*', *Philosophy of Science* 39 (1972), 99-101.

Semelaigne, R., 1930-1932, *Les pionniers de la psychiatrie française avant et après Pinel,* Baillère, Paris, 2 vols.

Sérieux, P. and Capgras, J., 1909, *Les folies raissonnantes: le délire d'interprétation,* Alcan, Paris.

One of the classical French works on paranoia, contemporary with Kraepelin.

Shakow, D., 1947, *The Nature of Deterioration in Schizophrenic Conditions.*

Nervous & Mental Disease Publishers, New York.
A cognitive approach to schizophrenia stressing defective attention as the primary defect. Reminiscent of P. Janet's concept of the consciousness as a field.

Shapiro, D., 1965, *Neurotic Styles* (Austen Riggs Center Monograph Series, No. 5) Basic Books, New York.
Clinical phenomenological description of exceptional clarity, compassion and comprehension of the diverse pathological styles – the obsessive, hysterical, depressive and paranoid.

Siegler, M. and Osmond, H., 1974, *Models of Madness, Models of Medicine,* Macmillan, New York and London.
This is an eminently readable and clear survey of the field. The authors are sane, sensible, and reasonable; they rightly commend explicit and clear statements of positions. Its scope is wider than the one we confine ourselves to, since it treats etiology and treatment, as well as prevention, on top of diagnosis. We disagree, however, with much of what they advocate. Though we agree that the doctor's authority be rather explicit than implicit, we would not show half of their readiness to tolerate it. And though we support their call for clear diagnosis we are not happy at all with the diagnostic tools that they refer to: the medical model that they support does not exist: it is not a model but a set of *desiderata* for one – which may not be as easy to find as they hope.
See also Paul H. Blaney's review of that book in *Psychiatry* 39 (1976), 98-100. We fully agree with the reviewer's assessment.

Siegler, M. Osmond, H., and Mann, H., 1969, 'Laing's Model of Madness', *Brit. J. Psychiat.* 115, 947-958, repr. in R. Boyers and R. Orrill (eds.), *Laing and Anti-Psychiatry,* Penguin, Harmondsworth, 1971.
Siegler *et al.* tried to put Laing's views in a form of models in a mock serious hostile review. It is an interesting attempt nonetheless. Yet, we believe, it does not cover all Laing's views and is unnecessarily scurrilous.

Silberer, H., 1909, 'Report on a Method of Eliciting and Observing Certain Symbolic Hallucination-Phenomena', pp. 195-207, in D. Rapaport (ed.), 1951, *Organization and Pathology of Thought,* Columbia University Press, New York.

Silberer, H., 1912, 'On Symbol-Formation', pp. 208-233, in D. Rapaport (ed.), 1951, *Organization and Pathology of Thought,* Columbia University Press, New York.
The celebrated Silberer experiments. He tried to visualize abstract concepts in a state of semi-sleep. It is essentially a study of how an abstract concept may 'become' a 'concrete' picture.

Simmel, G., 1955, *Conflict* (transl. by K. H. Wolff), The Free Press, New York.
See Chapter 3, note 18.

Slater, E., 1973 and 1975, 'The Psychiatrist in Search of a Science', *Brit. J. Psychiat.* 122, 625-636; and 126, 205-224.
A lucid and critical appraisal of the state of contemporary psychiatry, by one of the pillars of the profession.

Spielberger, C. (ed.), 1966, *Anxiety and Behaviour,* Academic Press, New York.

A collection of very careful studies by the captains in the field.

Spitzer, B. L., 1976, 'More on Pseudoscience in Science and the Case for Psychiatric Diagnosis', *Arch. Gen. Psychiat.* 33, 459-470.
See note 28 to Chapter 3 above.

Stanton, A. H. and Schwartz, M. S., 1954, *The Mental Hospital,* Tavistock, London.
See Chapter 3, note 28.

Stengel, E., 1959, 'Classification of Mental Disorders', *W.H.O.* 21, 60-663.
A survey.

Straus, E. and Zutt, J., (eds.), 1963 (with Sattes, H.,), *Die Wahnwelten (Endogene Psychosen),* Akademische Verlag, Frankfurt a/M.
Views of classical and of contemporary psychiatrists concerning delusions. Among the phenomenologists, note the works by Schneider, K., 'Zum Begriff des Wahns'; Straus, E., 'Die Ästhesiologie und ihre Bedeutung für das Verständnis der Halluzinationen'; Binswanger, L., 'Vom anthropologischen Sinn der Verstiegenheit'; Zutt, J., 'Vom ästetischen im Unterschied zum affektiven Erlebnisbereich'; Kulenkampff, C., 'Entbergung, Entgrenzung, Überwältigung als Weisen des Standverlustes. Zur Anthropologie der paranoiden Psychosen'.

Sullivan, H. S., 1940, *The Interpersonal Theory of Psychiatry,* Norton, New York, 1953.

Sullivan, H. S., 1940, *Conceptions of Modern Psychiatry,* Norton, New York, 1953; Tavistock, London, 1955.

Sullivan, H. S., 1956, *Clinical Studies in Psychiatry,* Norton, New York.
Widely known as the founder of the Interpersonal Relationships School in psychiatry. Sullivan's disciples began the succesful treatment of psychosis.

Szalita-Pemow, A. B., 'Remarks on Pathogenesis and Treatment of Schizophrenia', *Psychiatry* 14 (1951), 295-300.

Szalita-Pemow, A. B., 'Further Remarks on the Pathogenesis and Treatment of Schizophrenia', *Psychiatry* 15 (1952), 143-150.

Szalita, A. B., 'Regression and Perception in Psychotic States', *Psychiatry* 21 (1958), 53-63.
The author of these lucid papers follows the ideas of Freud, Sullivan and Melanie Klein. She notices some fundamental differences between psychosis and neurosis. In particular, whereas the neurotic represses incestuous, aggressive, and other guilt, and thus anxiety promoting feelings, the psychotic is easily able to bring them to the fore, but uses dissociation as a defense rather than repression. The author would, we think, take Laing's observation of a psychotic's distance from his body as one instance of dissociation. Likewise she notices that Freudian regression must be split into the return to an early emotional stage of the neurotic on the one hand and the descent to a lower functional level of the psychotic on the other. She concludes that psychotic obsessive thinking is not symbolic but a primary thought disorder.
All this, we think, is incorporated in our presentation too, and we would be glad to be viewed as going in her footsteps just as much as in those of Ey.

Szasz, T. S., 1961, *The Myth of Mental Illness, Foundations of a Theory of*

Personal Conduct, Hoeber-Harper, New York, new ed., 1974.
Mental illness is not a disease in the medical sense of the word; it is a particular state in human communication.

Szasz, T. S., 1963, *Law, Liberty and Psychiatry: An Inquiry into the Social Use of Mental Health Practices,* Macmillan, New York.
"In the most elementary sense to be at liberty – or to be free – means to be unfettered ... The impediment I want to consider here is restraint on persons exercised by psychiatrists by virtue of the power vested in them by law. For those oppressed by psychiatrists, liberty means freedom from psychiatric coercion." (p. 6.)

Szasz, T. S., 1970, *The Manufacture of Madness: A Comparative Study of the Inquisition and the Mental Health Movement,* Harper and Row, New York, 1970.
A repetition of Szasz' liberal thesis, stated more historically than before, and against a broader cultural background. His most mature book.

Szasz, T. S., 1970, *Ideology and Insanity,* Anchor/Doubleday, New York.

Vigotsky, L. S., 1930-1934,*Thought and Language,* Wiley, London & New York, 1962.
A classic. For Vigotsky, in contrast to Piaget, the child is first a social being and affirms his individuality only later. Vigotsky was a pioneer in the U.S.S.R. in the study of thinking in schizophrenia. E. Hanfmann and J. Kasanin (via Luria) continued his work in the U.S.A. *See* Appendix I above.

Waelder, Ra, 1951, 'The Structure of Paranoid Ideas: a Critical Survey of Various Theories', *Int. J. Psychoanal.* 32, 167-177.

Walshe, F. M. R., 1948, *Critical Studies in Neurology,* Livingstone, Edinburgh.
A Jacksonian view of modern neurology.

Weakland, J., 1956, *see* Bateson, G. *et al.* (1956).

Werner, H. 1948, *Comparative Psychology of Mental Development,* Science editions, New York, 1961.
Animal behaviour, ethnology, child psychology, and psychopathology viewed under the common denominator "primitive thinking". *See* note 7 to Appendix I above.

Werner, H. and Kaplan, B., 1963, *Symbol Formation: An Organismic-Developmental Approach to Language and the Expression of Thought,* Wiley, New-York.
Werner and Kaplan believe that the formation of a symbol is not an arbitrary affair; it is closely linked with the preferred object or situation. Thus, early speech tends to be onomatopoeic. Later in life, they say "distancing" takes place and symbols become more abstract except for regressive cases. *See* note 9 to Chapter 9 above.

Wheelis, A., 1958, *The Quest for Identity,* Norton, New York.
Wheelis describes the ego as including, in addition to Freud's space-time- and cause-co-ordination, a sense of one's own place in the co-ordinate system. Laing calls this "ontological security".

Whewell, W., 1840, *The Philosophy of the Inductive Sciences,* Parker, London, 1847.

Whewell, W., 1958, *Novum Organum Renovatum.*

This most celebrated philosopher is coming into his own again. A brilliant synthesis between Kant and Bacon: in order to see, we must first speculate; our speculations are usually false and should be empirically refuted; yet, with luck and ingenuity, we may hit upon a true speculation, which we fail to refute, and thus verify, and thus render scientific.

Wilden, A., 1968, *see* Lacan, J. (1966).

Wilson, R. N., *see* Leighton, A. H. (1957).

Wisdom, J. O., 1961, 'A Methodological Approach to the Problem of Hysteria', *Int. J. Psychoanal.* 42, 231.

Wisdom, J. O., 1962, 'Comparison and Development of the Psycho-Analytic Theories of Melancholia', *Int. J. Psychoanal.* 43, 113-132.

Wisdom, J. O., 1970, 'Freud and Melanie Klein: 'Psychology, Ontology and *Weltanschauung*', in Hanly, C. and Lazerowitz, M. (eds.), 1970, *Psychoanalysis and Philosophy,* International Universities Press, New York.

Wisdom systematized our knowledge of hysteria (1961) and of melancholia (1962) from a Kleinian viewpoint; he arrives at a clear demarcation between neurosis and psychosis (1970).

Wootton, B., 1959 (assisted by Seal, V. G., and Chambers, R.), *Social Science and Social Pathology,* Allen & Unwin, London.

"A jurist and social scientist, contend that contemporary psychiatry and social work are repressive welfare agencies. ... forensic psychiatrists perform moral functions in the guise of medical action". (Szasz, 1963). We are not sure that Barbara Wooton will class herself as a Szaszian anti-psychiatrist: but she is undoubtedly not only a sharp critic of current methodology in the social sciences, but also a strong defender of the individual's rights, when psychiatrists label him sick on moral or social grounds, not on medical grounds. The worst is, she claims, they do it quite innocently, being unaware of the overlapping of the moral and the medical.

Yap, P. M., 1974, *Comparative Psychiatry,* Toronto University Press, London & Toronto.

A synthesis between the general bio-psychological factor of mental illness common to all men, and the specific socioculturally tainted coloured syndromes.

Yellowlees, H., 1946, *To Define True Madness,* Penguin, Harmondsworth.

A remarkably free wheeling and undogmatic presentation, popular yet instructive for the profession.

Zilboorg, G., and Henry G., 1941, *A History of Medical Psychology,* Norton, New York.

Psycho-dynamically oriented.

Zutt, J., 1963, *see* Straus, E. and Zutt, J., 1963.

INDEX OF NAMES

SYNTHESE LIBRARY

Monographs on Epistemology, Logic, Methodology,
Philosophy of Science, Sociology of Science and of Knowledge, and on the
Mathematical Methods of Social and Behavioral Sciences

1. J. M. BOCHEŃSKI, *A Precis of Mathematical Logic*. 1959, X+100 pp.
2. P. L. GUIRAUD, *Problèmes et méthodes de la statistique linguistique*. 1960, VI+146 pp.
3. HANS FREUDENTHAL (ed.), *The Concept and the Role of the Model in Mathematics and Natural and Social Sciences, Proceedings of a Colloquium held at Utrecht, The Netherlands, January 1960*. 1961, VI+194 pp.
4. EVERT W. BETH, *Formal Methods. An Introduction to Symbolic Logic and the Study of Effective Operations in Arithmetic and Logic*. 1962, XIV+170 pp.
5. B. H. KAZEMIER and D. VUYSJE (eds.), *Logic and Language. Studies dedicated to Professor Rudolf Carnap on the Occasion of his Seventieth Birthday*. 1962, VI+256 pp.
6. MARX W. WARTOFSKY (ed.), *Proceedings of the Boston Colloquium for the Philosophy of Science, 1961-1962,* Boston Studies in the Philosophy of Science (ed. by Robert S. Cohen and Marx W. Wartofsky), Volume I. 1973, VIII+212 pp.
7. A. A. ZINOV'EV, *Philosophical Problems of Many-Valued Logic*. 1963, XIV+155 pp.
8. GEORGES GURVITCH, *The Spectrum of Social Time*. 1964, XXVI+152 pp.
9. PAUL LORENZEN, *Formal Logic*. 1965, VIII+123 pp.
10. ROBERT S. COHEN and MARX W. WARTOFSKY (eds.), *In Honor of Philipp Frank,* Boston Studies in the Philosophy of Science (ed. by Robert S. Cohen and Marx W. Wartofsky), Volume II. 1965, XXXIV+475 pp.
11. EVERT W. BETH, *Mathematical Thought. An Introduction to the Philosophy of Mathematics*. 1965, XII+208 pp.
12. EVERT W. BETH and JEAN PIAGET, *Mathematical Epistemology and Psychology*. 1966, XII+326 pp.
13. GUIDO KÜNG, *Ontology and the Logistic Analysis of Language. An Enquiry into the Contemporary Views on Universals*. 1967, XI+210 pp.
14. ROBERT S. COHEN and MARX W. WARTOFSKY (eds.), *Proceedings of the Boston Colloquium for the Philosophy of Science 1964-1966, in Memory of Norwood Russell Hanson,* Boston Studies in the Philosophy of Science (ed. by Robert S. Cohen and Marx W. Wartofsky), Volume III. 1967, XLIX+489 pp.

15. C. D. BROAD, *Induction, Probability, and Causation. Selected Papers.* 1968, XI+296 pp.
16. GÜNTHER PATZIG, *Aristotle's Theory of the Syllogism. A Logical-Philosophical Study of Book A of the Prior Analytics.* 1968, XVII+215 pp.
17. NICHOLAS RESCHER, *Topics in Philosophical Logic.* 1968, XIV+347 pp.
18. ROBERT S. COHEN and MARX W. WARTOFSKY (eds.), *Proceedings of the Boston Colloquium for the Philosophy of Science 1966-1968,* Boston Studies in the Philosophy of Science (ed. by Robert S. Cohen and Marx W. Wartofsky), Volume IV. 1969, VIII+537 pp.
19. ROBERT S. COHEN and MARX W. WARTOFSKY (eds.), *Proceedings of the Boston Colloquium for the Philosophy of Science 1966-1968,* Boston Studies in the Philosophy of Science (ed. by Robert S. Cohen and Marx W. Wartofsky), Volume V. 1969, VIII+482 pp.
20. J. W. DAVIS, D. J. HOCKNEY, and W. K. WILSON (eds.), *Philosophical Logic.* 1969, VIII+277 pp.
21. D. DAVIDSON and J. HINTIKKA (eds.), *Words and Objections: Essays on the Work of W. V. Quine.* 1969, VIII+366 pp.
22. PATRICK SUPPES, *Studies in the Methodology and Foundations of Science. Selected Papers from 1911 to 1969,* XII+473 pp.
23. JAAKKO HINTIKKA, *Models for Modalities. Selected Essays.* 1969, IX+220 pp.
24. NICHOLAS RESCHER *et al.* (eds.), *Essays in Honor of Carl G. Hempel. A Tribute on the Occasion of his Sixty-Fifth Birthday.* 1969, VII+272 pp.
25. P. V. TAVANEC (ed.), *Problems of the Logic of Scientific Knowledge.* 1969, XII+429 pp.
26. MARSHALL SWAIN (ed.), *Induction, Acceptance, and Rational Belief.* 1970, VII+232 pp.
27. ROBERT S. COHEN and RAYMOND J. SEEGER (eds.), *Ernst Mach; Physicist and Philosopher,* Boston Studies in the Philosophy of Science (ed. by Robert S. Cohen and Marx W. Wartofsky), Volume VI. 1970, VIII+295 pp.
28. JAAKKO HINTIKKA and PATRICK SUPPES, *Information and Inference.* 1970, X+336 pp.
29. KAREL LAMBERT, *Philosophical Problems in Logic. Some Recent Developments.* 1970, VII+176 pp.
30. ROLF A. EBERLE, *Nominalistic Systems.* 1970, IX+217 pp.
31. PAUL WEINGARTNER and GERHARD ZECHA (eds.), *Induction, Physics, and Ethics, Proceedings and Discussions of the 1968 Salzburg Colloquium in the Philosophy of Science.* 1970, X+382 pp.
32. EVERT W. BETH, *Aspects of Modern Logic.* 1970, XI+176 pp.
33. RISTO HILPINEN (ed.), *Deontic Logic: Introductory and Systematic Readings.* 1971, VII+182 pp.
34. JEAN-LOUIS KRIVINE, *Introduction to Axiomatic Set Theory.* 1971, VII+98 pp.
35. JOSEPH D. SNEED, *The Logical Structure of Mathematical Physics.* 1971, XV+311 pp.
36. CARL R. KORDIG, *The Justification of Scientific Change.* 1971, XIV+119 pp.
37. MILIČ ČAPEK, *Bergson and Modern Physics,* Boston Studies in the Philosophy of Science (ed. by Robert S. Cohen and Marx W. Wartofsky), Volume VII. 1971, XV+414 pp.

38. Norwood Russell Hanson, *What I do not Believe, and other Essays,* (ed. by Stephen Toulmin and Harry Woolf), 1971, XII+390 pp.
39. Roger C. Buck and Robert S. Cohen (eds.), *PSA 1970. In Memory of Rudolf Carnap.* Boston Studies in the Philosophy of Science (ed. by Robert S. Cohen and Marx W. Wartofsky), Volume VIII. 1971, LXVI+615 pp. Also available as paperback.
40. Donald Davidson and Gilbert Harman (eds.), *Semantics of Natural Language.* 1972. X+769 pp. Also available as paperback.
41. Yehoshua Bar-Hillel (ed.), *Pragmatics of Natural Languages.* 1971, VII+231 pp.
42. Sören Stenlund, *Combinators, λ-Terms and Proof Theory.* 1972, 184 pp.
43. Martin Strauss, *Modern Physics and Its Philosophy. Selected Papers in the Logic, History, and Philosophy of Science.* 1972, X+297 pp.
44. Mario Bunge, *Method, Model and Matter.* 1973, VII+196 pp.
45. Mario Bunge, *Philosophy of Physics.* 1973, IX+248 pp.
46. A. A. Zinov'ev, *Foundations of the Logical Theory of Scientific Knowledge (Complex Logic),* Boston Studies in the Philosophy of Science (ed. by Robert S. Cohen and Marx W. Wartofsky), Volume IX. Revised and enlarged English edition with an appendix, by G. A. Smirnov, E. A. Sidorenka, A. M. Fedina, and L. A. Bobrova. 1973, XXII+301 pp. Also available as paperback.
47. Ladislav Tondl, *Scientific Procedures,* Boston Studies in the Philosophy of Science (ed. by Robert S. Cohen and Marx W. Wartofsky), Volume X. 1973, XII+268 pp. Also available as paperback.
48. Norwood Russell Hanson, *Constellations and Conjectures,* (ed. by Willard C. Humphreys, Jr.), 1973, X+282 pp.
49. K. J. J. Hintikka, J. M. E. Moravcsik, and P. Suppes (eds.), *Approaches to Natural Language. Proceedings of the 1970 Stanford Workshop on Grammar and Semantics.* 1973, VIII+526 pp. Also available as paperback.
50. Mario Bunge (ed.), *Exact Philosophy – Problems, Tools, and Goals.* 1973, X+214 pp.
51. Radu J. Bogdan and Ilkka Niiniluoto (eds.), *Logic, Language, and Probability.* A selection of papers contributed to Sections IV, VI, and XI of the Fourth International Congress for Logic, Methodology, and Philosophy of Science, Bucharest, September 1971. 1973, X+323 pp.
52. Glenn Pearce and Patrick Maynard (eds.), *Conceptual Change.* 1973, XII+282 pp.
53. Ilkka Niiniluoto and Raimo Tuomela, *Theoretical Concepts and Hypothetico-Inductive Inference.* 1973, VII+264 pp.
54. Roland Fraïssé, *Course of Mathematical Logic* – Volume 1: *Relation and Logical Formula.* 1973, XVI+186 pp. Also available as paperback.
55. Adolf Grünbaum, *Philosophical Problems of Space and Time.* Second, enlarged edition, Boston Studies in the Philosophy of Science (ed. by Robert S. Cohen and Marx W. Wartofsky), Volume XII. 1973, XXIII+884 pp. Also available as paperback.
56. Patrick Suppes (ed.), *Space, Time, and Geometry.* 1973, XI+424 pp.
57: Hans Kelsen, *Essays in Legal and Moral Philosophy,* selected and introduced by Ota Weinberger. 1973, XXVIII+300 pp.
58. R. J. Seeger and Robert S. Cohen (eds.), *Philosophical Foundations of*

Science. Proceedings of an AAAS Program, 1969. Boston Studies in the Philosophy of Science (ed. by Robert S. Cohen and Marx W. Wartofsky), Volume XI. 1974, X+545 pp. Also available as paperback.

59. Robert S. Cohen and Marx W. Wartofsky (eds.), *Logical and Epistemological Studies in Contemporary Physics,* Boston Studies in the Philosophy of Science (ed. by Robert S. Cohen and Marx W. Wartofsky), Volume XIII. 1973, VIII+462 pp. Also available as paperback.
60. Robert S. Cohen and Marx W. Wartofsky (eds.), *Methodological and Historical Essays in the Natural and Social Sciences. Proceedings of the Boston Colloquium for the Philosophy of Science, 1969–1972,* Boston Studies in the Philosophy of Science (ed. by Robert S. Cohen and Marx W. Wartofsky), Volume XIV. 1974, VIII+ 405 pp. Also available as paperback.
61. Robert S. Cohen, J. J. Stachel and Marx W. Wartofsky (eds.), *For Dirk Struik. Scientific, Historical and Political Essays in Honor of Dirk J. Struik,* Boston Studies in the Philosophy of Science (ed. by Robert S. Cohen and Marx W. Wartofsky), Volume XV. 1974, XXVII+652 pp. Also available as paperback.
62. Kazimierz Ajdukiewicz, *Pragmatic Logic,* transl. from the Polish by Olgierd Wojtasiewicz. 1974, XV+460 pp.
63. Sören Stenlund (ed.), *Logical Theory and Semantic Analysis. Essays Dedicated to Stig Kanger on His Fiftieth Birthday.* 1974, V+217 pp.
64. Kenneth F. Schaffner and Robert S. Cohen (eds.), *Proceedings of the 1972 Biennial Meeting, Philosophy of Science Association,* Boston Studies in the Philosophy of Science (ed. by Robert S. Cohen and Marx W. Wartofsky). Volume XX. 1974, IX+444 pp. Also available as paperback.
65. Henry E. Kyburg, Jr., *The Logical Foundations of Statistical Inference.* 1974, IX+421 pp.
66. Marjorie Grene, *The Understanding of Nature: Essays in the Philosophy of Biology,* Boston Studies in the Philosophy of Science (ed. by Robert S. Cohen and Marx W. Wartofsky), Volume XXIII. 1974, XII+360 pp. Also available as paperback.
67. Jan M. Broekman, *Structuralism: Moscow, Prague, Paris.* 1974, IX+117 pp.
68. Norman Geschwind, *Selected Papers on Language and the Brain,* Boston Studies in the Philosophy of Science (ed. by Robert S. Cohen and Marx W. Wartofsky), Volume XVI. 1974, XII+549 pp. Also available as paperback.
69. Roland Fraïssé, *Course of Mathematical Logic* – Volume II: *Model Theory.* 1974, XIX+192 pp.
70. Andrzej Grzegorczyk, *An Outline of Mathematical Logic. Fundamental Results and Notions Explained with All Details.* 1974, X+596 pp.
71. Franz von Kutschera, *Philosophy of Language.* 1975, VII+305 pp.
72. Juha Manninen and Raimo Tuomela (eds.), *Essays on Explanation and Understanding. Studies in the Foundations of Humanities and Social Sciences.* 1976, VII+440 pp.
73. Jaakko Hintikka (ed.), *Rudolf Carnap, Logical Empiricist. Materials and Perspectives.* 1975, LXVIII+400 pp.
74. Milič Čapek (ed.), *The Concepts of Space and Time. Their Structure and Their Development.* Boston Studies in the Philosophy of Science (ed. by Robert S. Cohen and Marx W. Wartofsky), Volume XXII. 1976, LVI+570 pp. Also available as paperback.

75. JAAKKO HINTIKKA and UNTO REMES, *The Method of Analysis. Its Geometrical Origin and Its General Significance.* Boston Studies in the Philosophy of Science (ed. by Robert S. Cohen and Marx W. Wartofsky), Volume XXV. 1974, XVIII+144 pp. Also available as paperback.
76. JOHN EMERY MURDOCH and EDITH DUDLEY SYLLA, *The Cultural Context of Medieval Learning. Proceedings of the First International Colloquium on Philosophy, Science, and Theology in the Middle Ages – September 1973.* Boston Studies in the Philosophy of Science (ed. by Robert S. Cohen and Marx W. Wartofsky), Volume XXVI. 1975, X+566 pp. Also available as paperback.
77. STEFAN AMSTERDAMSKI, *Between Experience and Metaphysics. Philosophical Problems of the Evolution of Science.* Boston Studies in the Philosophy of Science (ed. by Robert S. Cohen and Marx W. Wartofsky), Volume XXXV. 1975 XVIII+193 pp. Also available as paperback.
78. PATRICK SUPPES (ed.), *Logic and Probability in Quantum Mechanics.* 1976, XV+541 pp.
80. JOSEPH AGASSI, *Science in Flux.* Boston Studies in the Philosophy of Science (ed. by Robert S. Cohen and Marx W. Wartofsky), Volume XXVIII. 1975, XXVI+553 pp. Also available as paperback.
81. SANDRA G. HARDING (ed.), *Can Theories Be Refuted? Essays on the Duhem-Quine Thesis.* 1976, XXI+318 pp. Also available as paperback.
84. MARJORIE GRENE and EVERETT MENDELSOHN (eds.), *Topics in the Philosophy of Biology.* Boston Studies in the Philosophy of Science (ed. by Robert S. Cohen and Marx W. Wartofsky), Volume XXVII. 1976, XIII+454 pp. Also available as paperback.
85. E. FISCHBEIN, *The Intuitive Sources of Probabilistic Thinking in Children.* 1975, XIII+204 pp.
86. ERNEST W. ADAMS, *The Logic of Conditionals. An Application of Probability to Deductive Logic.* 1975, XIII+156 pp.
89. A. KASHER (ed.), *Language in Focus: Foundations, Methods and Systems. Essays dedicated to Yehoshua Bar-Hillel.* Boston Studies in the Philosophy of Science (ed. by Robert S. Cohen and Marx W. Wartofsky), Volume XLIII. 1976, XXVIII+679 pp. Also available as paperback.
90. JAAKKO HINTIKKA, *The Intentions of Intentionality and Other New Models for Modalities.* 1975, XVIII+262 pp. Also available as paperback.
93. RADU J. BOGDAN, *Local Induction.* 1976, XIV+340 pp.
95. PETER MITTELSTAEDT, *Philosophical Problems of Modern Physics.* Boston Studies in the Philosophy of Science (ed. by Robert S. Cohen and Marx W. Wartofsky), Volume XVIII. 1976, X+211 pp. Also available as paperback.
96. GERALD HOLTON and WILLIAM BLANPIED (eds.), *Science and Its Public: The Changing Relationship.* Boston Studies in the Philosophy of Science (ed. by Robert S. Cohen and Marx W. Wartofsky), Volume XXXIII. 1976, XXV+289 pp. Also available as paperback.

SYNTHESE HISTORICAL LIBRARY

Texts and Studies
in the History of Logic and Philosophy

Editors:

1. M. T. BEONIO-BROCCHIERI FUMAGALLI, *The Logic of Abelard*. Translated from the Italian. 1969, IX + 101 pp.
2. GOTTFRIED WILHELM LEIBNIZ, *Philosophical Papers and Letters*. A selection translated and edited, with an introduction, by Leroy E. Loemker. 1969, XII + 736 pp.
3. ERNST MALLY, *Logische Schriften*, ed. by Karl Wolf and Paul Weingartner. 1971, X + 340 pp.
4. LEWIS WHITE BECK (ed.), *Proceedings of the Third International Kant Congress*. 1972, XI + 718 pp.
5. BERNARD BOLZANO, *Theory of Science*, ed. by Jan Berg. 1973, XV + 398 pp.
6. J. M. E. MORAVCSIK (ed.), *Patterns in Plato's Thought. Papers arising out of the 1971 West Coast Greek Philosophy Conference*. 1973, VIII + 212 pp.
7. NABIL SHEHABY, *The Propositional Logic of Avicenna: A Translation from al-Shifā: al-Qiyās*, with Introduction, Commentary and Glossary. 1973, XIII + 296 pp.
8. DESMOND PAUL HENRY, *Commentary on De Grammatico: The Historical-Logical Dimensions of a Dialogue of St. Anselm's*. 1974, IX + 345 pp.
9. JOHN CORCORAN, *Ancient Logic and Its Modern Interpretations*. 1974, X + 208 pp.
10. E. M. BARTH, *The Logic of the Articles in Traditional Philosophy*. 1974, XXVII + 533 pp.
11. JAAKKO HINTIKKA, *Knowledge and the Known. Historical Perspectives in Epistemology*. 1974, XII + 243 pp.
12. E. J. ASHWORTH, *Language and Logic in the Post-Medieval Period*. 1974, XIII + 304 pp.
13. ARISTOTLE, *The Nicomachean Ethics*. Translated with Commentaries and Glossary by Hypocrates G. Apostle. 1975, XXI + 372 pp.
14. R. M. DANCY, *Sense and Contradiction: A Study in Aristotle*. 1975, XII + 184 pp.
15. WILBUR RICHARD KNORR, *The Evolution of the Euclidean Elements. A Study of the Theory of Incommensurable Magnitudes and Its Significance for Early Greek Geometry*. 1975, IX + 374 pp.
16. AUGUSTINE, *De Dialectica*. Translated with the Introduction and Notes by B. Darrell Jackson. 1975, XI + 151 pp.